Practice in Physics

T. B. Akrill
G. A. G. Bennet
C. J. Millar

A companion to the single
volume text: *Physics*

Edward Arnold

First published 1979
by Edward Arnold (Publishers) Ltd
41 Bedford Square, London WC1B 3DQ

British Library Cataloguing in Publication Data

ISBN 0 7131 0338 8

A companion to the single volume text: *Physics*

Acknowledgements

The publishers would like to acknowledge the following for providing photographs:

T B Akrill: 22-21

Kodansha Ltd: 22-2

C J Millar: 12-41

Physics is Fun: Book 3 J T Jardine (Heineman Educational Books Ltd): 2-9

PSSC Physics, 1st ed (D C Heath & Co 1960): 1-5, 1-36, 3-49, 23-4

Waves D C Chaundy (Longman) 27-4

Typesetting by Parkway Group London and Abingdon
Printed in Great Britain by
Fletcher & Son, Ltd, Norwich

Contents

Preface

We have written this book as a companion volume to our textbook *Physics*. We have designed the two volumes as a single unit to support a complete advanced physics course. But this book can also be used in conjunction with any physics textbook to provide additional practice material. The student we have in mind is one who has recently tackled a science examination at 16 +, and is embarking on a more advanced course. The material is freshly thought out for use at this level, and the mathematical skill we demand is adjusted accordingly. The principles we have followed are fully set out in the preface to *Physics*.

Most of the questions and exercises are concerned with helping the student to become familiar with the basic concepts of the subject. Very few questions are of the sophistication that a student must expect in an examination at the end of such a course. We have deliberately not included questions from past examination papers, as the student will in any case be using those of his own board towards the end of his course. In this volume we provide scope for revision, but the main purpose of the book is, as its name implies, to provide practice material for the first four or five terms of the course.

The reader will notice that the chapter and section headings in this book are identical with those in *Physics*. We have provided practice material to cover each point of significance in the main text. In every section we start with the most basic applications of a new idea, and then endeavour to lead the student through to a full understanding of the concepts involved. We do include a certain number of questions which require thought, of course, but we avoid all trick questions. We do not want to puzzle the student. We would rather he gained the confidence in handling new material that comes from successful achievement. So where difficult lines of reasoning must be developed we try and help the student by the wording of the question (sometimes by direct hints), and by asking for necessary intermediate steps of reasoning.

Inevitably numerical exercises form a large part of this volume. But we also try to provide practice material for every aspect of the subject. We include practice in the drawing and interpretation of graphs, in the handling of laboratory data, photographs or tabulated material, as well as practice in making estimates.

This book, like its companion volume, is the product of our combined creative efforts in the past three years. We offer it to our fellow teachers and their students as the distillation of our collective experience.

Advice to the student

It is only by *doing* physics that you will come to understand it. Hence the need for a book of practice material like this. You will notice that the questions and exercises in this book exactly match, section by section, the material in our textbook *Physics*. You can study a section in the textbook. Then try it out by working through the questions that go with it. In this way you will quickly gain confidence in handling the new ideas you will be learning.

To help you check how you are getting on, there are answers to all numerical questions at the end of the book. For questions that have a verbal answer we sometimes give a few words to indicate the gist of what you should have said (but in such cases *you* should reckon to write a rather more extended explanation). We advise you to use an electronic calculator, preferably of the scientific kind that handles powers of 10 and trigonometric functions. But be careful to give every answer to the right number of significant figures.

You must not just blindly put down all the digits registered by the calculator. The simple rule (with occasional obvious exceptions) is that the number of significant figures in the answer should be the same as the *smallest* number of significant figures in any item of data of the problem. You may find that this rule is sometimes a bit crude, but it is good enough for you to adopt at this stage.

All scientific measurements and calculations are nowadays conducted using the quantities and units of the *Système International* (SI). The symbols used, and their meanings, are summarised for quick reference in the tables on page 115.

January 1979

T. B. Akrill
G. A. G. Bennet
C. J. Millar

1 Describing motion

Data $g = 9.81$ m s^{-2}

Measuring speed

1-1 The age of the Earth is about 10^{17}s and the time taken for light from this page to reach your eye is about 10^{-9} s. Produce a list of events which take times which you estimate to be of orders of magnitude between these two extremes, i.e. taking 10^{15} s, 10^{13} s etc.

1-2 A 100 metre track is marked out with an uncertainty in its length of $\pm$ 0.50 m, and a stop-watch used for timing a runner on the track has an uncertainty of $\pm$ 0.20 s. If the stopwatch records a time of 12.5 s for a particular runner, what is the runner's average speed?

Which of the measurements introduces the greatest uncertainty into the result, and what is this uncertainty?

1-3 A starting pistol is fired in front of a microphone and the sound is arranged to start a scaler-timer which counts in milliseconds. When the sound reaches another microphone 12.0 m away the scaler-timer is stopped. If the speed of sound in air is 340 m s^{-1} (a) calculate the time which would be recorded by the scaler-timer and (b) the uncertainty in this result.

1-4 The figure shows the results of three sets of measurements of the same physical quantity, e.g. the diameter of a piece of wire. Comment on the three sets which were taken with different measuring instruments, two micrometer screw gauges and a vernier caliper. Your comments should include words like *accurate, sensitive, precise*.

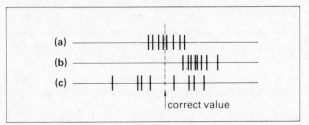

1-5 Measure the distances between the centres of the images of the ball in the photograph. The diameter of the ball shown was 50 mm. Use this to calculate the distances moved by the ball between successive exposures. If the exposures were made at intervals of 1/25 s, find the average speed of the ball for each of the distances you have calculated, and plot a speed-time graph.

1-6 The figure shows a piece of paper tape on which a ticker-timer has been making 50 dots each second; the tape was twice the size of the tape shown in the figure. Measure the lengths of successive five-space lengths of tape, starting from the left-hand edge of the tape, and calculate the average speed for each of these five-space lengths. Hence plot a graph of speed against time.

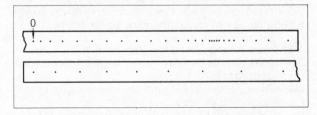

1-7 How would you use school laboratory apparatus to measure the speed of (a) a snail (b) a model train (c) an air rifle pellet?

Velocity, a vector quantity

1-8 What is the change of velocity when (a) $+ 6.0$ m s^{-1} becomes $+ 5$ m s^{-1} (b) $+ 6.0$ m s^{-1} becomes $- 15$ m s^{-1} (c) $+ 6.0$ m s^{-1} becomes $- 6.0$ m s^{-1} (d) 5.0 m s^{-1} east becomes 15 m s^{-1} west?

1-9 Estimate (a) your average velocity and (b) your average speed when you travel to school or work from your home.

1-10 The figure shows a particle moving in a circle ABCDA of radius 8.0 m at a constant speed. It completes one revolution in 5.0s. What is (a) its average *speed* for one revolution (b) its average *speed* from A to B (c) its average *velocity* for one complete revolution (d) its average *velocity* from A to C (e) its average *velocity* from A to B (f) its change of velocity from A to C (g) its change of velocity from A to B?

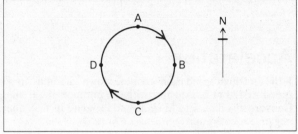

1-11 Use a scale diagram to add (a) a displacement of 5.6 m north to a displacement of 3.5 m NW (b) a velocity of 3.2 m s^{-1} SW to a velocity of 1.8 m s^{-1} N 20° W.
[Note: it helps to draw scale diagrams on graph paper.]

1-12 What is the total displacement of (a) a man who walks 5.0 m north and then 2.0 m south (b) a car which drives 100 m north and then 50 m east (c) a boy who runs once round a 400 m running track (d) a boy who runs halfway round a circular 400 m running track, starting at its most southerly point?

1-13 A man has a motor boat which he always drives at a speed of 2.5 m s^{-1}. He starts to cross a river which is 50 m wide, and which runs east-west. The whole river can be assumed to flow at a speed of 2.0 m s^{-1} west. If he starts from the south bank and aims due north (a) what will be his velocity (b) how long will it take him to reach the north bank (c) how far west of his starting point will he then be?

1-14 Refer to question **1-13**. If the man wants to reach the north bank directly opposite his starting point, he must aim the boat upstream.
(a) What angle should this direction make with the bank?
(b) What will be the velocity of the boat?
(c) How long will he take to reach the north bank?

1-15 The figure shows a body's initial velocity v_1 and final velocty v_2. Calculate its change of velocity.

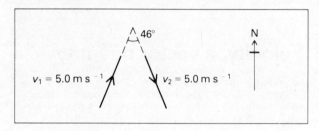

1-16 Use a scale diagram to subtract (a) a displacement of 5.0 m in a direction N 60° W from a displacement of 10.0 m in a direction N 20° E (b) a velocity of 6.5 m s^{-1} in a direction N 40° W from a velocity of 4.0 m s^{-1} in a direction S 30° W.

1-17 An aeroplane is found to be flying in a NW direction at 120 m s^{-1} when there is a steady 35 m s^{-1} SW wind. What would be the velocity of the aeroplane if the wind ceased?

Acceleration

1-18 The 'high speed train' can slow down smoothly from a speed of 120 m.p.h. to rest within a distance of 1.0 mile. Convert this data into SI units and estimate its maximum negative acceleration (1 mile ≃ 1.6 km).

1-19 The figure shows (full-size) a piece of ticker tape with three successive dots on it; the ticker-timer was making 50 dots each second. Measure the distances between the dots and deduce the acceleration (assumed to be constant) of the body to which the tape was attached.

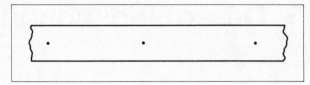

1-20 One type of aeroplane has a maximum acceleration on the ground of 3.4 m s^{-2}. What is the minimum length of runway needed if it is to reach its take-off speed of 110 m s^{-1} and how long a time will this take?

1-21 A man standing at the edge of a cliff leans over and throws a stone straight up into the air so that eventually it falls into the sea below him. If the stone leaves his hand with a speed of 12 m s^{-1} what is the stone's displacement after (a) 2.0 s (b) 3.0 s? [Choose the upward direction to be positive.]

1-22 An air track glider is placed on a linear air track which is slightly tilted. It is given a velocity of 1.5 m s^{-1} up the track. If its acceleration is 2.0 m s^{-2} down the track, find the time at which it is 1.0 m below its starting point.

1-23 A particle moves in a straight line. Its motion can be described as follows: $t = 0$, $v = 0$; $0 < t < 10$ s, $a = 4.0$ m s^{-2}; 10 s $< t < 20$ s, $a = -4.0$ m s^{-2}. Sketch the velocity-time graph and use it to find the change of displacement of the particle between $t = 0$ and $t = 20$ s.

1-24 A road test report gives the following data for a standing start acceleration test for a car.

t/s	0	5	10	15	20	25	30	35	40
v/m s^{-1}	0	14	24	30	34	37	39	40	40

Draw a velocity-time graph for the car (using graph paper), and hence find the displacement of the car when it has reached a speed of (a) 25 m s^{-1} (b) 35 m s^{-1}. [Keep your graph for question **1-25**.]

1-25 Refer to question **1-24**. Use the velocity-time graph to find (a) the acceleration of the car when its speed is 30 m s^{-1} (b) the maximum acceleration of the car.

1-26 The figure shows a velocity-time graph for a particle. Describe as fully as possible the motion it represents.

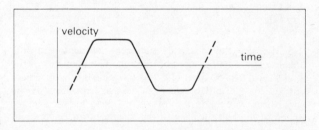

1-27 A man throws a ball straight up into the air and catches it again. Take the upward direction to be positive and sketch the velocity-time graph for the ball, assuming that air resistance is small enough to be neglected. How could you use your graph to find the height reached by the ball?

2

1-28 Draw a velocity-time graph for a sky diver who is released from rest and who falls through the air from a great height. Explain the shape of the graph.

1-29 Draw velocity-time graphs for the following situations. In (a) and (c) take the downward direction to be positive, and ignore air resistance. In (b) and (c) assume that some of the kinetic energy is converted into internal energy at each bounce.
(a) A tennis ball falling to the ground.
(b) A billiard ball striking a cushion at right angles, and bouncing backwards and forwards between opposite cushions.
(c) A tennis ball falling to the ground and bouncing several times.

1-30 Draw displacement-time, velocity-time and acceleration-time graphs for the following situations. Use the same time axes for all three graphs for each situation. You will probably find it easiest to begin with the velocity-time graph.
(a) An electrically-driven milk float moving from one house to another on a straight road.
(b) A ball, attached by an elastic cord to a fixed point on the ground, which is hit horizontally away from that point.

1-31 A boy tosses a coin in a stationary railway carriage in such a way that the coin rises and falls vertically, and is airborne for 0.40 s. Later he tosses the coin in an identical manner when (a) the train is moving at a constant velocity of 10 m s^{-1} (b) the train has a velocity of 10 m s^{-1} and a forward acceleration of 1.0 m s^{-2}. Where does the coin land on these occasions?

Free fall

1-32 How long does a ball take to fall a distance of (a) 1.0 m (b) 2.0 m? Why is the answer to (b) not twice the answer to (a)?

1-33 A steel ball is held by an electromagnet. When the current in the electromagnet is switched off the ball falls. As it leaves the electromagnet an electric circuit is broken and a scaler-timer starts. The ball falls a distance of 456 mm and strikes a trip-switch which stops the scaler-timer. The time recorded is 301 ms. What value does this experiment give for the free-fall acceleration g?

What sources of error are there likely to be in such an experiment?

1-34 A bullet is fired horizontally at a speed of 200 m s^{-1} at a target which is 100 m away. How far has the bullet fallen when it hits the target? What angle does its velocity make with the horizontal then? Explain whether air resistance would cause the bullet to fall a greater or lesser amount than the distance you give.

1-35 A body is projected with a speed of 25 m s^{-1} at an angle of 30° above the horizontal. Neglect air resistance, and calculate (a) the vertical resolved part of its velocity (b) the time taken to reach its highest point (c) the greatest height reached (d) the horizontal range of the body.

1-36 (a) Measure the horizontal distances of the second, fourth, sixth, etc., images of the ball from the left-hand edge of the photograph. Comment on your results.
(b) The horizontal lines in the photograph are actually 152 mm apart. Measure their separation on the photograph and hence calculate how much larger the actual distances are than the distances in the photograph. Hence find (in metres) the vertical distances s of the centres of the balls when the second, fourth, sixth, etc., flashes occurred. For vertical motion starting from rest $s = \frac{1}{2}gt^2$, so that $\sqrt{s} \propto t$. Draw a graph of $\sqrt{s}$ (on the y-axis) against time t (on the x-axis) to verify this. [Keep your graph for question **1-37**.]

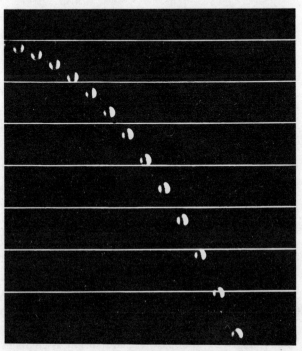

1-37 Refer to question **1-36**. If $s = \frac{1}{2}gt^2$, $\sqrt{s} = \sqrt{(\frac{1}{2}g)}t$, and the slope of the graph represents $\sqrt{(\frac{1}{2}g)}$. If the intervals between flashes were 1/30 s, measure the slope of the graph. Hence find a value for g.

1-38 A shell is fired from a gun on a hillside. Its initial velocity is 50 m s^{-1} at an angle of 53° above the horizontal. Ignore air resistance, and calculate the vertical and horizontal displacements of the shell at one-second intervals from $t = 0$ to $t = 10$ s. Hence draw, to scale, the path of the shell, and from your drawing measure the direction of its velocity when its horizontal displacement is 200 m.

1-39 The following readings were taken in an experiment with a simple pendulum whose length l could be varied:

l/m	number of oscillations	time/s
0.40	50	63
0.60	40	62
0.80	40	72
1.00	35	70
1.20	30	66

3

Given that the period T of the pendulum is related to l by $T = 2\pi\sqrt{(l/g)}$, plot these results in such a way as to obtain a linear graph, and deduce a value for g.

Explain why the experimenter counted at least 30 oscillations for each pendulum, and why he counted more oscillations for the shorter pendulums.

2 Momentum and force

Data $g = 9.81$ m s^{-2} = 9.81 N kg^{-2}

Mass and momentum

2-1 What is the momentum of (a) a boy of mass 50 kg running round a running track at a constant speed of 3.0 m s^{-1} at the moment when he is (i) facing north, (ii) facing south (b) a car of mass 800 kg moving east at a speed of 25 m s^{-1} (c) an oil tanker of mass 250 000 tonnes which is moving west at a speed of 20 m s^{-1}?

2-2 Two air-track gliders are at rest with a spring compressed between them: a thread tied to each stops them moving apart. When the thread is cut, one glider moves away with a speed of 0.32 m s^{-1}, and the other moves in the opposite direction with speed of 0.45 m s^{-1}. If the first glider has a mass of 0.40 kg, what is the mass of the other?

Could this experiment have been performed (a) on the Moon (b) deep in outer space? Would the result have been the same in each case?

2-3 Two trolleys A and B are placed on a laboratory bench with a spring compressed between them. When the spring is released the trolleys move apart. Both trolleys then hit, at the same instant, barriers placed some distance away: A travels 850 mm to its barrier, and B travels 670 mm to its barrier. What is the ratio (mass of A)/(mass of B)? Explain how you arrive at your answer, and state any assumptions you make.

2-4 (a) A girl of mass 60 kg steps, at a speed of 2 m s^{-1}, off a canoe, of mass 40 kg, onto a river bank. What happens to the canoe?
(b) A rugby player of mass 70 kg, running south at 6.0 m s^{-1}, tackles another player whose mass is 85 kg and who is running directly towards him at a speed of 4.0 m s^{-1}. If in the tackle they cling together, what will be their common velocity immediately after the tackle?

2-5 Body A of mass 3.0 kg has a velocity of + 4.0 m s^{-1} and collides head-on with body B, which has a mass of 2.0 kg and a velocity of − 2.0 m s^{-1}. After the collision the velocity of B is found to be + 1.5 m s^{-1}. Find the velocity of A.

Sketch a graph to show how the momenta of the two bodies varies with time before and after the collision.

2-6 An α-particle is emitted from a polonium nucleus with a speed of 1.800×10^7 m s^{-1}. If the relative masses of the α-particle and the remaining part of the nucleus are 4.002 and 212.0, find the recoil speed of the nucleus.

2-7 Two boys stand a few metres apart on a trolley which can move freely on a horizontal surface without any resisting force. One throws a heavy ball to the other, who catches it. Describe what happens to (a) the speed of the trolley (b) the position of the trolley, during this process.

2-8 A shell explodes into two parts (which may or may not be equal) at the moment when it is moving horizontally. Explain under what circumstances the two parts can (a) move in the same direction (b) move in opposite directions (c) move away on paths making an angle of 90° with each other. In each case give a diagram showing the paths of the two parts after the explosion.

2-9 The photograph shows the result of an air-track collision between two gliders. The upper trace shows the positions of the straws (fixed to the gliders) before the collision, and the lower track shows the positions of the straws after the collision. Make measurements on the photograph to find the speeds (in arbitrary units, e.g. mm per flash) of the gliders before and after the collision (on this occasion there is no need to know the scale of the photograph, since we are concerned only with the relative sizes of the speeds). The mass of the left-hand glider is 0.20 kg, and that of the right-hand glider is 0.30 kg. Bearing in mind the limited precision with which you were able to measure the distances, do you think that this experiment supports the fact that momentum is conserved in all collisions?

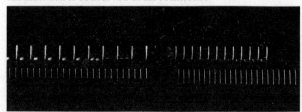

2-10 A bullet of mass 10 g is fired into a block of wood of mass 200 g and becomes embedded in it. If the speed of the bullet was 500 m s^{-1}, what is the speed of the block immediately afterwards?

2-11 Discuss in what way momentum is conserved when (a) a train accelerates from rest (b) a lump of plasticine falls to the ground and stays there (c) a ball falls to the ground and bounces up again.

2-12 Write a few sentences to explain clearly to someone what is meant by the *mass* of a body.

2-13 Standing passengers in a bus may fall over if the bus stops quickly. Which way do they fall, and why?

Forces

2-14 The figures shows (a) a man holding a suitcase (b) a man sitting on a chair and leaning his elbows on a table and (c) a man leaning against a wall. Draw a free-body diagram for each man. Mark the forces acting on the men, label the forces, and describe them in a phrase like '*W* is the pull of the Earth on the man'.

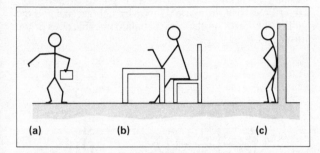

(a)　　　　(b)　　　　(c)

2-15 A force *P* is 20 N in a direction N 60° E. What is the resolved part of *P* in the following directions: (a) north (b) east (c) N 30° E?

2-16 Discuss whether the tension is the same throughout (a) two wires (whose mass is small enough to be neglected), one of which has twice the diameter of the other, which are joined end to end, hung vertically, and used to support a load (b) a heavy chain hanging vertically.

2-17 Describe three situations (as different as possible) where a body is moving with constant velocity. For each situation (a) draw a free-body diagram for the body, giving each force a symbol (b) write a phrase to describe each force, e.g. '*X* is the push of the air on the car' (c) say what you can about the sizes of the forces you have marked.

2-18 A block of wood of weight 6.0 N slides at constant speed down a slope which makes an angle of 20° with the horizontal.
(a) Is it in equilibrium?
(b) By resolving the forces perpendicular to and parallel with the slope calculate the size of the perpendicular contact force and the frictional force.

2-19 A framed picture of weight 50 N is to be hung on a wall, using a piece of string. The ends of the string are tied to two points, 0.60 m apart on the same horizontal level, on the back of the picture. Draw a free-body diagram for the picture, and find the tension in the string if it is (a) 1.0 m long (b) 0.66 m long.

2-20 One man uses a rope to pull a tractor horizontally: he pulls due north with a force of 450 N. Another man pulls horizontally with a force of 370 N in a direction N 50° E. Use a scale diagram to find the sum of these two forces.

2-21 A pendulum bob of weight *W* is attached to a thread of length *l* and hung from a rigid support. Another thread is attached to the bob: this thread is pulled sideways with a horizontal force *P* so that the bob moves sideways. The thread then makes an angle θ with the vertical, and the *horizontal* displacement of the bob is *d*. Show that (a) $\tan\theta = P/W$ (b) $\sin\theta = d/l$ and hence shows that the pull *P* is approximately proportional to *d*, if *d* is small compared with *l*.

2-22 Find a rod of circular cross-section (e.g. a felt-tip pen), a piece of string about 400 mm long, and an object with a mass of about 1 kg. Tie the string to the mass and wrap the string, somewhere near its middle point, round the rod as shown in the figure. Hold the rod steady, and try to raise or lower the object. The wind the string *twice* round the rod and repeat the experiment. Repeat with three, four, etc., loops. Account for the difference in the tensions in the string above and below the rod.

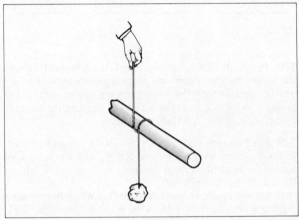

Newton's second law

2-23 The figure shows how the momentum of a body varies with time. Estimate the force acting on the body at (a) $t = 1.0$ s (b) $t = 2.0$ s.

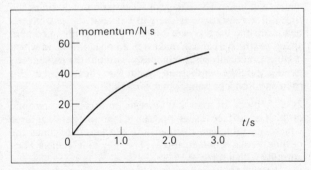

2-24 Why do two stones of *different* weights (e.g. 10 N and 20 N) fall with the *same* acceleration in a vacuum?

2-25 One girl throws a ball to another girl. Is it possible to state the direction of the ball's acceleration while it is in the air? [Assume that air resistance is negligible.]

2-26 A lift of mass 1200 kg is being pulled vertically upwards at a steady speed of (a) 1.0 m s⁻¹, (b) 2.0 m s⁻¹. What is the tension in the cable in each case?

2-27 The two pieces of ticker-tape shown in the figure were attached to the same body on two different occasions. What is the ratio of the forces which were exerted on the body on those occasions?

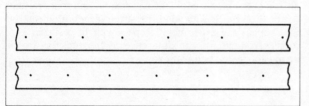

2-28 In one 10-minute interval during the Apollo 11 flight to the Moon the spacecraft's speed decreased from 5374 m s⁻¹ to 5102 m s⁻¹ (with the rocket motors not in use). If the mass of the spacecraft was 4.4×10^4 kg, find the average force on the spacecraft during this time.

2-29 (a) Two men push a car of mass 800 kg to get it started. If each pushes with a force of 300 N, and the resistance forces are equivalent to an opposing force of 160 N, what is the acceleration of the car?
(b) A tractor pulls a log of mass 2000 kg. When the tractor is pulling with a force of 1300 N the acceleration of the log is 0.050 m s⁻²: what resistance force does the ground exert on the log?

2-30 You wish to perform experiments using trolleys of mass 800 g on a runway, and you tilt the runway to compensate for frictional forces. At an angle of 6.0° the trolley accelerates down the slope at 0.50 m s⁻². Draw a free-body diagram for the trolley, and calculate the angle you should use to compensate correctly for frictional forces.

2-31 It is considered safe to design seat belts (for cars) which will give a passenger an acceleration of 20 times the free-fall acceleration. If the gap between the passenger's head and the windscreen is 0.60 m, what is the maximum speed at which a car can make a head-on collision, in which it stops practically instantaneously, without the passenger's head hitting the windscreen? What force does the seat belt then exert on a passenger of mass 70 kg?

2-32 A block of mass 2.0 kg rests on a rough horizontal table. The surface is such that the maximum frictional force which the table can exert on the block is 0.30 times the perpendicular contact force. The block is pulled horizontally with a force of 15 N.
(a) What is the acceleration of the block?
(b) What would be the acceleration of the block if its mass were doubled?

2-33 Refer to question **2-32**. Calculate the acceleration of the 2.0 kg block if the pull of 15 N made an angle of 20° with the horizontal. Why is this answer greater than the answer to question **2-32**(a)?

2-34 The table gives the results of a standing-start acceleration test for a car of mass 1100 kg. Draw a graph of its speed v against time t, and estimate the resultant force acting on the car when its speed was (a) 15 m s⁻¹ (b) 25 m s⁻¹.

t/s	0	2	4	6	8	10	12
v/m s⁻¹	0	9	15	19.5	23	26	29

2-35 The figure shows free-body diagrams for (a) a car accelerating along a straight horizontal road, and (b) a man pushing a packing case across a rough horizontal floor. Describe each of the marked forces by a phrase like 'P is the pull of the Earth on the car'. Estimate possible sizes of the forces.

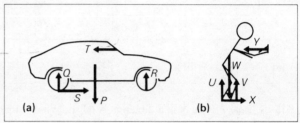

(a)   (b)

2-36 When one car tows another, why is it sensible (a) to keep the tow rope taut (b) for the towing car's acceleration to be small (c) for the rope to be made of a material like nylon which stretches a considerable amount when it is pulled?

2-37 A pendulum bob of mass 50 g hangs by a thread from the roof of a railway carriage. Describe and explain, using free-body diagrams for the pendulum bob, what happens to the bob when the train is (a) accelerating forwards (b) moving at constant velocity (c) slowing down to rest.
For (a) make any possible calculation if the acceleration is 0.80 m s⁻².

2-38 The figure shows a block on a horizontal frictionless table. A thread attached to it runs horizontally to a pulley at the edge of the table, passes over the pulley, and supports a load of mass 1.0 kg. The acceleration of both the block and the load is 2.0 m s⁻².

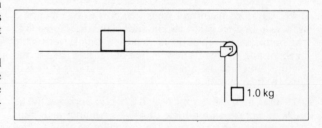

(a) Draw free-body diagrams for (i) the block, (ii) the load, labelling as T the pull of the thread on each of the bodies.
(b) Use the free-body diagram for the load to find the size of T.
(c) Now use the free-body diagram for the block to find the mass of the block.

2-39 In the figure the thin line shows the weight of a parachutist as he falls: his weight is constant. The thicker line shows the size of the air resistance force on him (and his parachute, when open) as he falls. He opens his parachute at time t_1, after falling freely. Explain the shape of this graph, and draw a graph to show the variation of his acceleration with time.

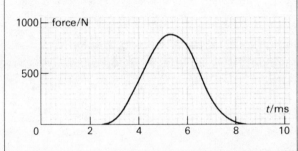

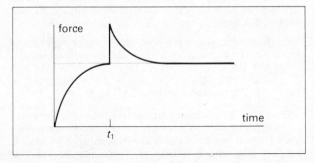

2-40 A boy stands on a diving board and holds at arm's length a spring balance which supports a load of 2.0 kg. He steps off the diving board and falls vertically into the pool. Can you say anything about the spring balance reading while he is falling through the air? Explain your answer.

2-41 A man stands still with both feet on some bathroom scales. Describe what happens to the reading on the scales when, separately (a) he presses downwards with his hands on a rail fixed to the wall (b) he lifts up one knee, sharply (c) he stands still on one leg.

2-42 What is the impulse of the forces in the following situations: (a) a man pulls a garden roller eastwards with a horizontal force of 300 N for 10 s (b) a stone of weight 20 N moving vertically downwards for 5.0 s (c) a stone of weight 20 N moving vertically upwards for 5.0 s?

2-43 A tennis ball of mass 60 g is moving horizontally, at right angles to the net, with a speed of 20 m s^{-1}. A player hits it straight back so that it leaves his racket with a speed of 25 m s^{-1}. What is (a) the change of momentum of the ball (b) the impulse of the force which the racket exerts on the ball?

2-44 A boy catches a cricket ball of mass 160 g which is moving at a speed of 20 m s^{-1}. Find the force which he must exert to stop it in (a) 0.10 s (b) 0.50 s. Describe how he can vary the time in this way, and explain the advantage of lengthening the time in which the ball is stopped.
Describe two other situations (as different as possible from this one) in which care is taken to lengthen the time in which an object is stopped.

2-45 The figure shows how the push of a tennis racket on a tennis ball of mass 60 g varies with time. Estimate the change of momentum of the ball, and, if it was initially at rest, its speed now.
What is the maximum acceleration of the ball?

2-46 A ring of mass 0.40 kg can slide freely on a frictionless horizontal rail. The ring starts from rest and is pulled for 3.0 s with a steady force of 0.10 N which makes an angle of 25° with the rail. How fast is the ring then moving?

2-47 Calculate the average force exerted by a golf club on a golf ball of mass 46 g, if the ball leaves the club at a speed of 80 m s^{-1} and the contact between club and ball lasts for 0.50 ms. State an object which would have a weight approximately equal to this force. Sketch a graph to show how the force might vary with time.

2-48 A railway truck of mass 10 tonnes is seen to be moving at a speed of 2.0 m s^{-1} to the right. After 3.0 s it collides with another truck of mass 30 tonnes moving at a speed of 1.0 m s^{-1} to the left. After the collision, which takes 1.0 s, the second truck is moving at 0.20 m s^{-1} to the left.
(a) What is the velocity of the first truck after the collision?
(b) Draw a graph of momentum against time for each truck, using the same axes, for the period $t = 0$ to $t = 7.0$ s. Assume that in the collision the momentum of each truck varies uniformly with time.
(c) What is the significance of the slopes of these graphs? Calculate the slopes for the period $t = 3.0$ s to $t = 4.0$ s.

2-49 Describe, giving reasons, how you would drive a car if there were a nearly-full bucket of water on the floor.

2-50 Why is a rubber-headed hammer not as good, for driving a nail into a block of wood, as a hammer with a head made of metal, even if both hammers have the same mass? [In your answer you should consider the ideas of *momentum, rate of change of momentum,* and *force.*]

Animal and vehicle propulsion

2-51 The figure shows (a) a man pulling a suitcase across a table and (b) and (c) free-body diagrams for the suitcase and the man respectively. Describe each of the forces in a phrase like '*A* is the pull of the Earth on the suitcase'. Do any of the forces have the same size because of Newton's third law?

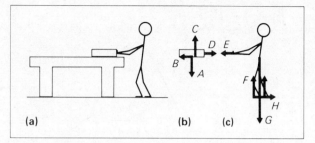

(a)　(b)　(c)

2-52 In question **2-51** several of the forces you have marked on the free-body diagrams are caused by the Earth (or by the ground, which forms part of the Earth). Draw a free-body diagram of the Earth for this situation, label the forces, and explain which of the forces you have marked in question **2-51** are equal, because of Newton's third law, to forces you have marked on the Earth in this question.

2-53 A man stands still on the surface of the Earth. Draw free-body diagrams for (a) the man (b) the Earth, marking and labelling the forces acting on these bodies. State which pairs of forces are equal because of Newton's first law, and which pairs are equal because of Newton's third law.

2-54 An articulated lorry consists of a tractor of mass 4.0 tonnes and a trailer of mass 26 tonnes. Ignoring all resistive forces, calculate, when the lorry has an acceleration of 0.20 m s^{-2}, (a) the push of the road on the driving wheels of the tractor (b) the pull of the tractor on the trailer.

2-55 Draw separate free-body diagrams for the tractor and trailer in question **2-54**. Which pairs of forces are equal because of Newton's third law?

2-56 A man of mass 65 kg stands on a weighing machine in a lift which has a downward acceleration of 3.0 m s^{-2}. What is the reading (in N) on the weighing machine? Make it clear at what stage you need to use Newton's third law. Explain why the answer does not depend on the direction in which the lift is moving.

2-57 A sprinter of mass 60 kg reaches his top speed of 12 m s^{-1} in the first 15 m of his run.
(a) Calculate his acceleration during this process.
(b) Hence calculate the horizontal force (assumed to be constant) which the ground has been exerting on him.
(c) What size force has he been exerting on the ground? Explain.
(d) What effect does this force have on the ground?

2-58 A packing case of mass 50 kg rests on a rough horizontal floor. The surface is such that the *maximum* frictional force which the floor can exert on the case is 0.40 times the perpendicular contact force. Describe what happens when a man pushes horizontally on the packing case with a force of (a) 98 N (b) 196 N (c) 294 N.

2-59 Describe some of the difficulties of living in a world where there were no frictional forces.

Jets and rockets

2-60 A wind of speed 10 m s^{-1} blows horizontally and at right angles to a cricket sight screen which measures 4.0 m by 10 m. Estimate the force which the wind exerts on the sight screen, explaining any assumptions you make. Density of air = 1.3 kg m^{-3}.

2-61 In heavy rain a depth of 30 mm of water might fall in one hour. If the rain falls vertically and its terminal speed is 8.0 m s^{-1}, calculate the average force on a flat roof of area 25 m^2 due to the *falling* rain (assume that no water lies on the roof). The density of water is 1000 kg m^{-3}.

2-62 A toy rocket whose initial mass is 80 g contains water and compressed air. When a lever is released the air pushes the water out of a nozzle of cross-sectional area 10 mm^2. If the density of water is 1000 kg m^{-3}, what must be the speed of the emerging water if lift-off is to be achieved?

What is the initial acceleration of the rocket if the exhaust speed is 10 m s^{-1}? Discuss how the acceleration varies subsequently.

2-63 The table gives some data about three rockets. Complete it.

rocket	thrust/N	mass flow/kg s^{-1}	exhaust speed/m s^{-1}
V2	2.5×10^5	180	
Atlas	1.8×10^6		1800
Saturn V		1.4×10^4	2400

2-64 A capsule containing 5.0 g of liquid carbon dioxide at a high pressure is attached to a 'dynamics' trolley which is placed on a friction-compensated runway in a laboratory. The end of the capsule is then pierced so that the carbon dioxide emerges rapidly, and the trolley accelerates in the opposite direction. Assuming that the emission of carbon dioxide occurs at a constant rate of 3.0 s, find the exhaust speed of the carbon dioxide, if the acceleration of the trolley (of mass 0.80 kg) is found to be a constant 0.25 m s^{-2}.

In practice how would you expect the acceleration of the trolley to vary with time?

2-65 If 70 kg of air passes through a jet engine each second, and the speed of the air is increased by 600 m s^{-1}, calculate the forward push on an aircraft fitted with four such engines.

As always, when a force pushes a body, some other body must be doing the pushing. In this situation what is it that is pushing?

3 Energy and its conservation

Data $g = 9.81\ \text{N kg}^{-1} = 9.81\ \text{m s}^{-2}$
$c = 3.00 \times 10^8\ \text{m s}^{-1}$
$1\ \text{MeV} = 1.6 \times 10^{-13}\ \text{J}$

Work and power

3-1 Draw a free-body diagram for a block of mass 6.0 kg which is being pulled across a rough horizontal table by a horizontal force of 30 N. The maximum frictional force which the table can exert on the block is 20 N. Find the work done by each of the forces in a horizontal displacement of 0.80 m.

3-2 A simple pendulum consists of a thread of length 700 mm and a bob of mass 60 g. The bob is pulled to one side until its *vertical* height above its lowest position is 20 mm, and is then released. What is the size of the pull of the Earth on the bob? Draw a free-body diagram for the bob when it is moving, and calculate the work done by each of the forces on the bob while (a) it moves to its lowest point (b) it moves from its lowest point to its extreme position on the other side.

3-3 If a man can exert a force of 400 N on the right-hand end of the lever shown in the figure, what mass can be supported on the other end?

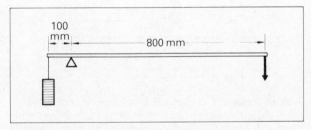

If he uses this force to push the right-hand end down 50 mm, how much work does he do? How much work is done by the other end of the lever on the load?

Would you describe this simple *machine* as a work-multiplier or a force-multiplier?

3-4 How much work is done on the Moon by the pull of the Earth when the Moon makes one complete orbit round the Earth? [Pull of Earth on Moon = 2.0×10^{20} N, radius of Moon's orbit = 3.8×10^8 m.]

3-5 To measure the work which can be done by a motor a *band brake* is sometimes used, as shown in the figure. The two spring balances measure the tensions in the rope on either side of the pulley, and the difference between the readings gives the friction force which the rope exerts on the pulley.
(a) Write down an expression for the work done by the motor in one revolution.

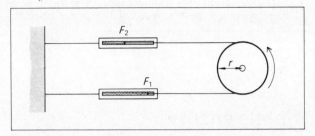

(b) If the spring balance readings are 9.5 N and 1.0 N when the motor is turning at 1200 revolutions per minute, and the radius of the pulley is 50 mm, what is the power of the motor?

3-6 A man does 20 press-ups (i.e. lying on his front, he straightens his arms to lift his shoulders from the ground) in 50 s. Estimate the power of those muscles which he is using.

3-7 A spring of constant stiffness 40 N m^{-1} is fixed at one end. A man pulls the other end horizontally. How much work does he do in stretching the spring from its unstretched position (a) 0.20 m (b) a further 0.20 m? Explain why the answer to (b) is not the same as the answer to (a).

3-8 The figure shows a force-extension graph for a wire which breaks at B. How much work has to be done on the wire to break it? How much work would have been needed if the wire had broken at A?

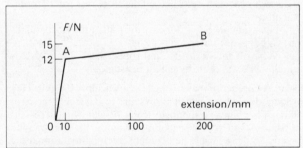

3-9 Corresponding values of force and extension for a particular rubber band are shown in the table.

force/N	0	0.25	0.50	0.75	1.0	2.0	3.0	4.0	5.0
extension/mm	0	20	40	50	55	64	66	68	70

Plot a graph of force (on the y-axis) against extension (on the x-axis). How much work is done in stretching the rubber band 70 mm?

3-10 The table shows how the Moon's gravitational pull F on a particular spacecraft varies with distance r from the centre of the Moon (radius 1.74×10^6 m).

$r/10^6$ m	1.74	2.0	2.4	2.8	3.2
$F/10^4$ N	6.5	4.9	3.4	2.5	1.9

Estimate the work done by this force as the spacecraft rises from the Moon's surface to a distance of 3.2×10^6 m from the centre of the Moon.

3-11 A motor drives a pulley which lifts a box of mass 5.0 kg at a steady speed of 2.0 m s^{-1}. What is the power output of the motor?

3-12 The power of the electric motor of a locomotive pulling a train at a constant speed of 50 m s^{-1} is 2.5 MW. What is then the total resistance force on the train?

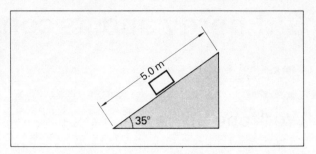

Kinetic energy

3-13 Estimate the kinetic energy of
(a) a tennis ball which has just been served. Tennis balls have a mass of 60 g, and the serving speed might be 30 m s^{-1}.
(b) a rifle bullet. Some have a mass of 10 g, and the speed might be 300 m s^{-1}.
(c) a man sprinting. The world record for a 100-metre run is less than 10 s.
(d) a loaded articulated lorry on a motorway. Its mass might be 40 tonnes, and 70 m.p.h. is roughly equal to 32 m s^{-1}.

3-14 (a) A car of mass 800 kg reaches a speed of 20 m s^{-1} in a distance of 100 m, as a result of internal forces doing work on the car. What constant external force would have been needed to do the same amount of work?
(b) If its motor is then switched off, and frictional forces equivalent to a constant horizontal force of 200 N act on it, in what distance will it stop?

3-15 An object of mass 300 g falls from rest for 2.5 m in a vacuum.
(a) How much work is done by the pull of the Earth on it?
(b) What is its final speed?
(c) When it fell in air its final speed was 6.8 m s^{-1}. How much work was done by the air resistance?

3-16 A climber of mass 70 kg falls vertically off a cliff. He is attached to a rope which allows him to fall freely for 20 m. Then it becomes taut, but stretches, bringing him to rest in a further 4.0 m. An energy-flow diagram for this accident would show (a) while he is falling freely, gravitational p.e. being converted into k.e. and (b) while he is being slowed down, the k.e., and some more g.p.e., being converted into elastic p.e. Draw this energy-flow diagram, attaching numerical values to the boxes for g.p.e. and k.e. Hence find the elastic p.e. stored in the rope, and the average force exerted on him by the rope.

3-17 A pendulum bob of mass 30 g is drawn to one side so that it is raised a *vertical* height of 35 mm, and released. How fast is it moving when it passes through its lowest point? Does this speed depend on the mass of the bob? Explain.

3-18 The figure shows a block of mass 4.0 kg sliding down a ramp of length 5.0 m. The frictional force is a constant 15 N. constant 15 N.
(a) How much work is done by the weight of the block?
(b) How much work is done by the frictional force?
(c) If the block was given a speed of 1.0 m s^{-1} at the top of the ramp, what is its speed at the bottom?

(d) If the block were required to slide down the ramp at constant speed, how much work would have to be done by the frictional force?
(e) *Hence* calculate the frictional force needed.

3-19 The figure shows part of a roller coaster at a fairground. If a car of mass 250 kg has a speed of 2.0 m s^{-1} at A, what is the maximum average force of the frictional and air resistance forces which may act on the car if it is to pass over the hump at C?

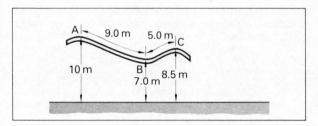

3-20 In a game of cricket a fielder throws a cricket ball of mass 160 g from the boundary to the wicket. It leaves his hand with a speed of 20 m s^{-1}. How fast is it moving when it has risen a vertical height of 5.0 m? Does the answer depend on the angle to the horizontal at which the ball is thrown?

3-21 The compression spring in a pin-ball machine has a stiffness of 100 N m^{-1}. The spring is compressed by a distance of 40 mm and when released fires a ball of mass 30 g. With what speed does it leave the spring?

3-22 Suppose that an α-particle (of mass 6.7×10^{-27} kg) starts at rest at a distance of 5.0×10^{-14} m from the centre of a particular nucleus. Estimate (a) its kinetic energy (b) its speed when it is 20×10^{-14}m from the centre of the nucleus, if the repelling force on it varies with distance from the centre of the nucleus as shown in the table.

distance/10^{-14}m	5.0	7.0	10	15	20
force/N	16.6	8.5	4.2	1.8	1.0

3-23 A spring of stiffness 20 N m^{-1} is hung vertically, with a lump of mass 0.50 kg fixed to its lower end.
(a) Show that the extension e of the spring when the mass is in equilibrium is 0.25 m.
(b) The mass is supported until the spring is just not stretched,

and is then released. How much work is done by the pull of the Earth on the lump while it falls the distance e?

(c) How much work is done by the pull of the spring on the lump while it falls the distance e?

(d) How much kinetic energy does the lump then have?

(e) What is the speed of the lump then?

(f) If the lump had been supported by hand and lowered gently so that it came to rest after moving down the distance e, how much work would have been done by the push of the hand in that distance?

(g) How would the size of the push of the hand have varied?

3-24 Refer to chapter 2, question **2-6**. What percentage of the released energy is carried away by the α-particle?

Potential energy

3-25 A squash ball of mass 25 g is allowed to fall from a height of 1.0 m onto a hard surface, and it bounces up to a height of 0.20 m.

(a) What is its initial gravitational p.e.?

(b) What is its final gravitational p.e.?

(c) How do you account for the difference?

3-26 How does a diver use a spring board to increase his or her height of jump? [An important feature of the process is the elastic potential energy stored in the spring board.]

3-27 Refer to the photograph which accompanies question **1-36** in chapter 1. A measurement of the distance between the centres of any two images, divided by the flash interval (1/30 s) will give the speed of the ball at the mid-point of that distance.

(a) Remembering to convert distances in the photograph to distances in the actual situation (the horizontal lines are actually 152 mm apart), find the speeds of the ball at each mid-point between images, and the height of this point above the bottom edge of the photograph. You should try to estimate the distances to the nearest 0.1 mm in the photograph.

(b) The mass of the ball is 0.10 kg. Calculate the k.e. and gravitational p.e. (taking the bottom edge of the photograph as the zero level) of the ball at each mid-point.

(c) Draw graphs of k.e. and g.p.e. against time, and also a graph of k.e. plus g.p.e. against time—all three graphs on the same axes.

3-28 A man attaches a load of mass 5.0 kg to the lower end of a vertical steel wire, supporting the load as he does so until the whole weight of the load is carried by the wire. The extension of the wire is then 4.0 mm. What is (a) the loss of gravitational p.e. of the load (b) the gain of elastic p.e. of the wire?

Why are these answers different?

3-29 The figure shows an energy-flow diagram. Suggest a process to which this might refer, giving numerical information where possible.

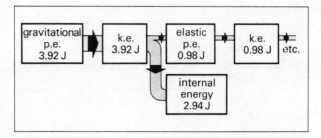

3-30 The figure shows a catapult which consists of two posts A and B, and a rubber band. When the band is placed round A and B it is just not stretched. The middle point of the rubber band is then pulled back to X (AX = XB = 90 mm); the force needed to hold the rubber band in this position is 5.0 N. What is the tension in the rubber band? Calculate the elastic p.e. now stored in it.

A pellet of mass 3.0 g is then placed at X, and the band is released, pushing the pellet to Y. Calculate the speed of the pellet when it loses contact with the rubber band at Y.

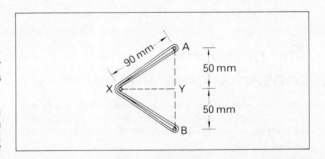

3-31 A lump of weight 4.0 N is held stationary at the bottom of a vertical spring of stiffness $20\ \text{N m}^{-1}$; the spring is not stretched. The lump is then released. When it has fallen 0.20 m, find the changes in (a) the gravitational potential energy of the lump (b) the elastic potential energy of the spring, (c) the kinetic energy of the lump.

The lump passes through this position. When it has fallen a total distance of 0.40 m find the change (from the initial values) in (d) the gravitational potential energy of the lump (e) the elastic potential energy of the spring. What can you deduce from these figures?

3-32 A trampolinist falls vertically from a height of 3.0 m: on the rebound he rises to a height of 3.5 m. Discuss the energy changes quantitatively and draw an energy-flow diagram for the process.

3-33 A catapult is used to push a glider of mass 0.20 kg along an air track. The greater the distance s by which the elastic in the catapult is pulled back, the faster the glider moves. A vertical card of length 100 mm fixed to the glider can interrupt the light beam, and the time t of interruption can be measured. The table gives corresponding values of s and t.

s/mm	10	16	24	29	37
t/s	1.25	0.77	0.53	0.43	0.33

(a) Find the speed v and hence the kinetic energy E_k of the glider for each value of s.

(b) The k.e. E_k should be equal to the elastic potential energy E_p stored in the catapult, and $E_p = \frac{1}{2} ks^2$, where k is the stiffness of the elastic. Draw a graph of E_p (on the y-axis) against s^2 (on the x-axis). Explain why the slope of the graph $= \frac{1}{2} k$, and find the value of k.

3-34 The tables (a) and (b) give corresponding values of gravitational potential energy W and distance r for two bodies, each in a different kind of gravitational field. Draw graphs of W (on the y-axis) against r (on the x-axis).

(a)
r/m	0	100	200	300	400	500
W/kJ	0	1.6	3.2	4.9	6.5	8.1

(b)
$r/10^8$ m	1	2	3	4	5
$W/10^9$ J	-6.3	-3.1	-2.1	-1.6	-1.3

The slope of such a graph gives a measure of the force acting on the body. What is the size of the force in (a)? In (b) the force varies in size. Find the force for $r = 2.0 \times 10^8$ m and $r = 4.0 \times 10^8$ m, and suggest a way in which the force might vary with distance (e.g. $F \propto 1/r$, $F \propto 1/r^2$, etc.).

The principle of conservation of energy

3-35 The table gives the rate at which energy is used by a typical human being in some common activities. Estimate (a) the energy you use in a day (b) the heating power of a class of 15 pupils.

activity	power/W	activity	power/W
sleeping	80	typewriting	160
lying down (awake)	90	walking (2 m s^{-1})	250
sitting still	120	running fast	700
standing	140	rowing in a race	1400

3-36 A motor car (internal combustion) engine is said to be 20% efficient, i.e., 80% of the chemical energy in the fuel is converted into useless internal energy. List the parts of the car where this conversion occurs. What happens to the 20% which is 'usefully' converted?

3-37 What does it mean to say that the muscles of the human body are 25% efficient?

3-38 A man doing heavy manual labour (e.g. digging a trench) converts energy at a rate of about 1500 W. *If* he could keep up this rate of working for a working day of 8 hours, how much work would he have done? Electrical energy is available at a cost of about 3p per kW h. What would be the cost of the energy equal to the work done by the man? Does your answer help to explain the improvement in the standard of living in developed countries in the past 200 years?

3-39 Gas and oil can be burnt in houses with an efficiency of about 85%. When their chemical energy is converted to electrical energy in power stations the efficiency of this process is typically 35%. Comment on the following points of view:

(a) electrical energy should be used only when it is necessary, e.g. in electric motors: gas and oil should be burnt in homes and factories for space heating.

(b) the use of gas and oil should be reserved for the chemical industry to use for manufacturing purposes (e.g. to make plastic materials).

3-40 The radiation received from the Sun at the Earth's surface at the latitude of Great Britain is typically about 600 W m^{-2} (in the absence of cloud). What area of solar panel would be needed to replace a power station of output 2.0 GW, if the solar panels used could convert solar radiation to electrical energy at an efficiency of 20%? What percentage is this area of the total area of the British Isles (which is 3×10^{11} m^2)? If the total power station capacity (1979) is about 80 GW, what percentage of the surface of Great Britain would be covered by solar panels if all the power stations were to be replaced? Comment on your answer.

3-41 There is a possibility of creating a *tidal barrage* on the River Severn. Water from the incoming tide would be trapped behind the barrage; it would then be released through turbines and therefore made to do work. Estimate the average power output from a power station using such turbines, given that there are two tides each day, that the average height of the tide is 10 m, that the area of the water behind the barrage is 70 km^2, and that the overall efficiency of the conversion process is 80%. The density of water is 1000 kg m^{-3}.

3-42 The graph shows the demand for electrical energy in a particular region during a 24-hour period. Estimate the mean demand rate during this period. It would obviously be desirable for the power station to be able to work at a constant rate, equal to this demand rate. It is therefore proposed to construct a power station to work at this mean demand rate: then, during periods when the availablity of energy is greater than the demand, the energy should be used to pump water up to a reservoir higher than the power station. It would be released to help drive the power station turbines when demand is greater than the mean demand rate. Assuming an efficiency of 80% in the pumping process, estimate the mass of water which would be needed, if the reservoir were 300 m above the power station.

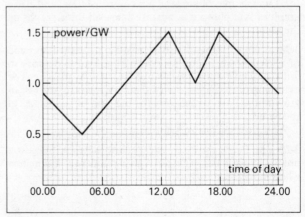

Collisions

3-43 A railway truck of mass 15 tonnes which has a velocity of 5.0 m s^{-1} north collides with a truck of mass 10 tonnes which has a velocity of 2.0 m s^{-1} south. They couple together on impact. Find their velocity after the collision, and the kinetic energy converted into other forms of energy.

3-44 Refer to the photograph which accompanied question 2-9 in chapter 2. Is that an elastic collision? Explain.

3-45 An air-track glider of mass 200 g, moving at 0.30 m s^{-1}, makes an elastic collision with a stationary glider of mass (a) 200 g (b) 400 g. What are the velocities of the two gliders after the collision?

3-46 A bullet of mass 10 g is fired into a block of wood of mass 4.0 kg which is supported by vertical threads. After the bullet has become embedded in the block, the block rises (vertically) 50 mm.
(a) What was the speed of the bullet?
(b) How much internal energy was produced?

3-47 Part (a) of the figure shows an air puck A of mass 100 g sliding with negligible frictional forces towards a stationary puck B of mass 200 g. Part (b) shows (not to scale) the velocities of the pucks after the collision. Draw a scale diagram (it helps to use graph paper) to add together the momenta of A and B after the collision. Verify that in the collision momentum has been conserved.

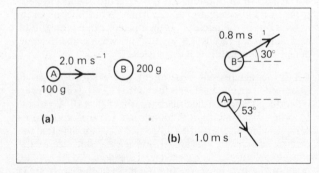

(a)

(b) 1.0 m s^{-1}

3-48 A Newton's cradle consists of five steel balls suspended by threads so that the balls lie in a horizontal line, almost touching each other. When one ball is pulled back and released so that it strikes head-on the row of the remaining four balls, the ball stops on impact, and the ball at the far end of the row moves off with the velocity which the first ball had, the other four now being stationary. Show that this is the result which should be expected if the collisions between the balls are elastic.

What do you expect to happen if two balls are pulled back together and then released?

[You can try this for yourself, using five coins (all the same) placed in a line on a polished table top. Place the coins between two rulers so that they stay in a straight line.]

3-49 The photograph shows a collision between a large ball, of mass 200 g, and a small ball. The large ball entered the

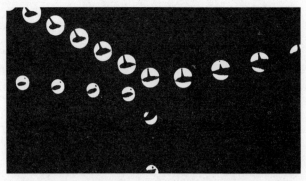

photograph from the right, and the small ball from the left. Draw scale diagrams to show the change of velocity of each ball during the collision. Hence find the mass of the small ball.

3-50 A car of mass 1000 kg which has a velocity of 10 m s^{-1} east collides at a crossroads with a car of mass 800 kg which has a velocity of 5.0 m s^{-1} north. If the cars lock together after the collision, find their initial velocity afterwards. How much kinetic energy is converted in the collision?

3-51 A particle of mass m emerges from a nucleus whose (remaining) mass is M, where $M > m$. Which particle has the larger kinetic energy?

3-52 In some situations an electron and a gas atom can collide elastically. Suppose the atom has a mass which is 1.0 $\times$ 10^5 times greater than that of the electron. Find the velocities of the electron and the gas atom after a collision in which (a) an electron of speed u hits, head-on, a stationary gas atom (b) a gas atom of speed u hits, head-on, a stationary electron. (You will find it sensible to make some approximations).

For (a) find the kinetic energy of the electron after the collision.

High-energy physics

3-53 How fast must a particle be moving for its mass to increase by (a) 1% (b) 10% (c) 50%? Express your answers as fractions of the speed of light c.

3-54 In lists of fundamental particles the units of mass are often given as MeV (1 MeV = 1.6 $\times$ 10^{-13} J). For example:

particle	rest mass/MeV
electron	0.512
proton	938

Express these masses in kilograms.

3-55 When one mole of hydrogen molecules reacts with half a mole of oxygen molecules in the chemical reaction $H_2 + \frac{1}{2}O_2 = H_2O$ the internal energy produced is 2.8 $\times$ 10^5 J. Express the loss in mass (a) in kilograms (b) as a fraction of the total mass of the original materials. (One mole of hydrogen molecules has a mass of 2.0 g; one mole of oxygen molecules has a mass of 32 g.)

3-56 One possible nuclear fusion reaction in the interior of the Sun involves the fusion of four hydrogen atoms (mass $1.007\,825\ m_u$) into one helium atom (mass $4.002\,603\ m_u$). If $m_u = 1.660\,43 \times 10^{-27}$ kg find the loss in mass (a) in kilograms (b) as a fraction of the mass of the original hydrogen (c) in MeV.

3-57 In a certain experiment electrons (of rest mass 9.11×10^{-31} kg) were accelerated to a very high speed and then allowed to travel at constant velocity for a distance of 5.00 m. The time t taken to travel this distance was recorded for different amounts of work W done (by electrical forces) in accelerating the electrons.

$W/10^{-14}$ J	0.50	1.00	2.00	3.00	4.00	5.00	6.00
$t/10^{-8}$ s	4.99	3.67	2.80	2.45	2.25	2.13	2.04

(a) Calculate the speed v_p which the electron could be predicted (using k.e. $= \frac{1}{2}mv^2$) to have when those amounts of work were done on it, and plot v_p^2 (on the y-axis) against W (on the x-axis). Since $W = \frac{1}{2}mv_p^2$ this graph should be a straight line of slope $2/m$.
(b) Calculate the actual speed v_a of the electron from the above measurements of time t. Plot the values of v_a, squared, on the same graph. The slope of this graph decreases. What does this suggest about the mass of the electron at very high speeds?

4 Structure of matter

Data Avogadro constant $L = 6.02 \times 10^{23}$ mol^{-1}
mass of proton $\simeq$ mass of neutron $= 1.66 \times 10^{-27}$ kg
mass of electron $= 9.11 \times 10^{-31}$ kg
unified atomic mass unit $m_u = 1.66 \times 10^{-27}$ kg
values of A_r: silver, 108; iron, 55.8
values of M_r: hydrogen, 2; helium, 4; nitrogen, 28; oxygen, 32

Atoms and molecules

4-1 A scientist working in the early nineteenth century might have made the following measurements on different compounds of nitrogen and oxygen:

	N/parts by 'weight'	O/parts by 'weight'
compound X	65.6	150
compound Y	380	217
compound Z	228	261

(a) He took the relative atomic mass (i.e. what he would have called the 'atomic weight') of oxygen to be 16. How many parts by weight of nitrogen combine with 16 parts by weight of oxygen in each of the compounds?
(b) Is it possible that the atomic weight of nitrogen is 7? If so, write out the simplest chemical formula of each of the compounds.
(c) Is it possible that the atomic weight of nitrogen is 14? If so, write out the simplest chemical formula of each of the compounds.
(d) On this evidence alone would you choose 7 or 14 to be the atomic weight of nitrogen? Why?

4-2 2.0 cm^3 of stearic acid are mixed with 18 cm^3 of alcohol, and 1.0 cm^3 of this solution is then added to 49 cm^3 of alcohol. This mixture is poured into a burette, and it is found that 5.0 cm^3 produce 65 drops. When one of these drops is gently lowered on to a water surface the drop spreads into a circular patch of diameter 290 mm. It may be assumed that the alcohol dissolves in the water. Calculate (a) the volume of stearic acid in the drop (b) the area of the circular patch and (c) the thickness of the layer. Molecules of stearic acid are known to consist of a 'backbone' of 18 carbon atoms. If when on the water surface they form a layer which is just one molecule thick, with the molecular backbone vertical (d) estimate the diameter of a carbon atom.

4-3 The diameters of most atoms are between 1×10^{-10} m and 5×10^{-10} m. Roughly how many atoms thick is (a) a piece of gold foil, of thickness about 5×10^{-7} m (b) a strand of the copper wire (diameter 0.10 mm) used to make multi-strand flexible electrical cable?

4-4 An atom of uranium-238 has 92 protons and 146 neutrons in its nucleus, and 92 electrons. What fraction of its mass is not in the nucleus? If the nucleus can be thought of as a sphere of radius 7.4×10^{-15} m, what is the average density of the material of the nucleus?

4-5 In an electrolysis experiment where hydrogen gas is liberated at an electrode it is found that 1.000 g of hydrogen is produced when an electric charge of 9.578×10^4 C passes. In other experiments it is found that 1.000 g of electrons carry an electric charge of 1.759×10^8 C. Assuming that the electric charge on a hydrogen ion is the same size as the charge on an electron, what can be deduced about the mass of the hydrogen ion and the mass of the electron? Is the assumption reasonable?

4-6 The atomic number Z of carbon is 6; the neutron numbers N of its three common isotopes are 6, 7 and 8. What are (a) the mass numbers A of these isotopes (b) the atomic mass of the most massive isotope?

4-7 The three common isotopes of silicon have mass numbers of 28, 29 and 30. If the percentage abundances are 92.2%, 4.7% and 3.1% respectively, find the relative atomic mass of silicon.

4-8 In one type of mass spectrometer positive ions of an element are all given the same kinetic energy and are then moving horizontally. They enter an evacuated horizontal tube, and while in the tube the only force which acts on them is the pull of the Earth on them. The time taken for them to reach the end of the tube is measured, and it is found that when potassium ions are used the ions arrive at two different lengths of time, corresponding to the isotopes of potassium which have mass numbers of 39 and 41.

(a) Explain which kind of ion arrives first. If the initial k.e. of the ions is 5.00×10^{-18} J, and the length of the tube is 0.600 m, what was (b) the initial speed of each type of ion and (c) the time of flight of each type of ion?

4-9 Find the mass of (a) one mole of electrons (b) 10 moles of oxygen molecules (c) 20 moles of oxygen atoms.

4-10 How many moles of (a) silver atoms are there in a silver coin of mass 10 g (b) iron atoms are there in a one-kilogram mass of iron?

4-11 Which of the following masses of gas (the molecular formula is given in brackets) contains the greatest number of molecules, and what is that number: (a) 20 g of hydrogen (H_2) (b) 20 g of helium (He) (c) 200 g of nitrogen (N_2)?

4-12 Find the relative molecular mass of palmitic acid ($C_{16}H_{32}O_2$), and use the value of L to calculate the mass of one molecule.

4-13 The relative atomic mass A_r of gold is 197, and its density in the solid state is 1.93×10^4 kg m^{-3}.
(a) How many atoms of gold are there in a volume of 1.00 m^3?
(b) What is the approximate 'diameter' of a gold atom?
(c) For aluminium $A_r = 26.9$, $\rho = 2.70 \times 10^3$ kg m^{-3}: repeat the calculations for aluminium. Comment on your answers.

4-14 Sodium chloride forms a cubical crystal in which all the layers are the same distance apart. Draw yourself a sketch of the way in which the ions of sodium and chlorine are arranged alternately in the crystal.
(a) If the layers of ions are 2.81×10^{-10} m apart, what is the volume of a 'cell' which contains one ion of sodium and one ion of chlorine?
(b) If the density of crystalline sodium chloride is 2.16×10^3 kg m^{-3}, what is the mass of this cell?
(c) The relative molecular mass of NaCl is 58.5: hence calculate a value for the Avogadro constant.

4-15 At some time in the past the following statements have been thought to be largely or wholly true. To what extent should they now be modified?
(a) Atoms can neither be created nor destroyed.
(b) All the atoms of any one element are identical.
(c) Atoms are far too small to be seen.
(d) Even in the smallest sample of an element the number of atoms is so great that it is quite impossible to know how many there are.

Atomic forces

4-16 Sketch a diagram of the atomic orbitals to show how (a) a molecule of lithium fluoride (LiF) is formed by ionic bonding (b) a molecule of ammonia (NH_3) is formed by covalent bonding. (Atomic numbers of these elements are H 1, Li 3, N 7, F 9.)

4-17 What are the four most common types of bonding?

Explain the mechanism of each, and give an example of a material formed by each type.

4-18 How is it possible for electrically *neutral* atoms to attract or repel each other?

4-19 The figure shows how the force of repulsion, and the force of attraction, between adjacent atoms in a crystal of a metallic element varies with their separation r. Copy the graphs, and add to it a graph to show the resultant force acting on each atom. [Note: at the separation r. the sizes of the attraction and repulsion forces are the same.] What is the significance of the separation r?

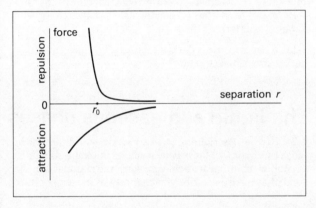

Using the graph of resultant force, explain (a) what happens when forces are exerted on the solid to (i) compress it, and (ii) stretch it (b) why the crystal obeys Hooke's law for small stresses.

4-20 Neighbouring ions in a crystal of sodium chloride are 2.8×10^{-10} m apart. If the repelling force F_r on an ion is proportional to $1/r^{10}$ and the attracting force F_a is proportional to $1/r^2$, and each is 5.1×10^{-9} N when an ion is in equilibrium, find the values of F_r and F_a, and the resultant force, (a) when $r = 2.4 \times 10^{-10}$ m and (b) when $r = 3.2 \times 10^{-10}$ m. State whether the resultant force is repulsive or attractive, and comment on their sizes.

4-21 The atoms in a particular kind of crystal have potential energies E when they have a separation r as shown in the table:

$r/10^{-10}$ m	2.0	2.2	2.5	2.8	3.0
$E/10^{-19}$ J	13.1	-4.2	-11.6	-12.7	-12.4

$r/10^{-10}$ m	3.5	4.0	4.5	5.0
$E/10^{-19}$ J	-11.2	-9.9	-8.9	-8.0

Draw a graph of $E/10^{-19}$ J against $r/10^{-10}$ m.
(a) The p.e. is a minimum when $r = 2.8 \times 10^{-10}$ m. What is the significance of this separation?
(b) Using a larger scale for the p.e. (e.g. 20 mm = 1.0×10^{-19} J) draw again the p.e. against separation graph for $r = 2.5 \times 10^{-10}$ m to $r = 3.0 \times 10^{-10}$ m. If the atoms have a k.e. of 0.2×10^{-19} J when $r = 2.8 \times 10^{-10}$ m, what are their maximum and minimum separations in their oscillation?

(c) Looking at the general shape of the graph, are the amplitudes of oscillation on both sides of the equilibrium position exactly the same? What does your answer predict about the behaviour of the solid material when it is heated?
(d) Draw tangents to the original graph at $r = 2.5 \times 10^{-10}$ m and $r = 2.8 \times 10^{-10}$ m and find their slope. What does the slope represent?

The solid phase

4-22 Why is it nowadays common to find the following items made of a plastic material (previously-used material in brackets): (a) washing-up bowls (enamelled iron) (b) rain-water pipes (cast iron) (c) electric light switches (brass with ceramic inserts)?

4-23 What evidence is there for thinking that many materials have a crystalline structure?

The liquid and gaseous phases

4-24 Give a description, as to someone with little knowledge of Physics, of the way in which we think the atoms of a crystalline material are behaving when the material is in the solid phase. Also describe what happens as the temperature of the material is raised until it melts, and eventually evaporates.

4-25 What would be the effect on the Brownian motion of ash particles in air of (a) using smaller ash particles (b) warming the air?

4-26 When liquid bromine is released at the bottom of a wide vertical tube containing air, it evaporates to form a brown gas, but the colour spreads only slowly up the tube: it takes perhaps 10 minutes to spread upwards for about 0.25 m. Yet the bromine molecules have speeds of about 200 m s^{-1}. Explain the slowness of the movement of the colour.

4-27 The mean free path of air molecules at s.t.p. is about 10^{-7} m: their speed is about 500 m s^{-1}. It can be shown statistically that when a molecule takes N 'steps' in random directions it reaches, on average, a distance of $\sqrt{N}$ steps from its starting point. How long would it be likely to take one molecule to move from one end to the other of a room 5.0 m long, if there are no air currents in the room?

4-28 The density of liquid oxygen at 90 K (its n.b.p.) is 1.14×10^3 kg m^{-3}, and the density of gaseous oxygen at the same temperature is 3.79 kg m^{-3}. If the 'diameter' of an oxygen molecule is d, the volume it occupies when in liquid oxygen is, roughly, d^3. In terms of d^3, how much space does a molecule take up when it is in gaseous oxygen at 90 K? What would be, roughly, the side of a cube of this volume?

5 Performance of materials

Data $g = 9.81$ N kg^{-1}

Materials in tension

5-1 What is the tensile stress in (a) a supporting cable of a suspension bridge which has a diameter of 40 mm and which pulls up the roadway with a force of 30 000 N (b) a nylon fishing line of diameter 0.35 mm which a fish is pulling with a force of 15 N (c) a tow rope of diameter 6.0 mm which is giving a car of mass 800 kg an acceleration of 0.40 m s^{-2} (other horizontal forces on the car are negligible)?

5-2 What is the tensile strain when (a) a copper wire of length 2.0 m has an extension of 0.10 mm (b) a rubber band of length 50 mm is stretched to a length of 150 mm?

5-3 A wire has one end fixed to a ceiling, and hangs vertically. At the other end is hanging a mass of 6.0 kg. What is the tension in the wire (a) at its lower end (b) at its upper end? What assumption must you make to be able to answer (b)?
Suppose the wire were replaced by a rope which had a mass of 1.0 kg. What is the tension in the rope (c) at its lower end (d) at its upper end (e) at its mid-point?

5-4 A rectangular strip of polythene is 0.10 mm thick and 10 mm wide (and several centimetres long). When it is stretched it deforms so that the ends still have the same width and thickness, but there is a central section which is still 0.10 mm thick but only 5.0 mm wide. If the force with which each end is being pulled is then 50 N, find (a) the tensions (b) the tensile stresses, in the wide and narrow parts of the strip.

5-5 Two wires of the same material but of different lengths and diameters are joined end-to-end and hung vertically and support a load. Which of the following quantities are the same for both wires: tensile force, tensile stress, strain, extension?
The same wires are now hung vertically side-by-side with their ends joined together. Which of the following quantities are now the same: extension, strain, tensile stress, tensile force?

5-6 Suppose a man of mass 80 kg jumps off a wall, which is 4.0 m above the ground.
(a) What is his speed on reaching the ground?
(b) If he takes 50 ms to stop when he reaches the ground, what is the average force exerted on him by the ground?

(c) Estimate the maximum force exerted on him by the ground.

(d) If the cross-sectional area of one of his tibia (the main bone in his leg) has a minimum value of 3.0×10^{-4} m^2, what is the maximum compressive stress in each of his tibia during the landing?

(e) This value is quite a large fraction of the maximum compressive stress of bone (about 16×10^7 Pa). What would you advise the man to do when he reaches the ground?

5-7 What is the minimum radius of a nylon fishing line which is required to lift a fish of mass 5.0 kg vertically at a steady speed? If the fish struggles (and therefore accelerates) the force may be increased by a factor of 10. What minimum radius is now required? Ultimate tensile stress of nylon filament = 60 MPa.

5-8 A lift of mass 3000 kg is supported by a steel cable of diameter 20 mm. The maximum acceleration of the lift is 2.5 m s^{-2}. Calculate (a) the maximum tension in the cable (b) the maximum stress in the cable (c) the maximum strain in the cable. [E for steel = 200 GPa.]

5-9 A copper wire of length 1.2 m and cross-sectional area 0.10 mm^2 is hung vertically; for copper, $E = 130$ GPa. A steadily increasing force is applied to its lower end to stretch it; when the force has reached a value of 10 N (a) what is the stress in the wire (b) what is the strain in the wire (c) what is the extension of the wire (d) how much elastic potential energy is stored in the wire?

What differences would there be in the energy conversions if, instead, a load of mass 1.0 kg were fixed to the end of the unstretched wire, and released?

5-10 A copper wire of cross-sectional area 0.10 mm^2 is hung vertically and loaded steadily. A tungsten wire of the same length but with a cross-sectional area of 0.15 mm^2 is also hung vertically, and equal steadily increasing forces are applied to both wires. The Young modulus E and ultimate tensile stress σ_u of copper and tungsten are given in the table:

	E/GPa	σ_u/MPa
copper	130	220
tungsten	410	120

Which of the wires (a) will break first (b) will have the larger extension at that time?

5-11 A garage intends to tow a lorry, using a steel wire. Explain to what extent it need consider each of the following factors: the Young modulus of the steel, the ultimate tensile stress of the steel, the stress in the steel when it ceases to behave elastically, the length of the wire, the cross-sectional area of the wire?

5-12 The figure shows idealised forms of the force-extension curve for specimens (of the same shape) of high tensile steel and mild steel up to the point where each specimen breaks.
(a) Which is the *stronger* material?
(b) By considering, in each case, the area between the graph and the extension-axis, find the work which must be done to break each specimen.
(c) Which is the *tougher* material?

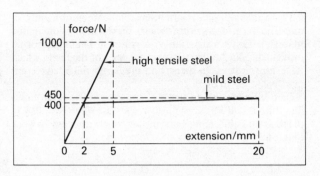

5-13 The Young moduli of copper and aluminium are 130 GPa and 70 GPa respectively. A bar of the same length and diameter is made from each material. Which would be the easier to bend? Explain.

5-14 Two students A and B perform experiments on copper wire. Both wires have the same length, but the diameter of B's is twice the diameter of A's. Explain which of the following statements are true:
(a) when they apply the same stress they get the same strain
(b) when they apply the same load A gets four times the strain that Y gets
(c) when they apply the same load A gets four times the extension that Y gets.

5-15 Wires of many materials form a neck just before they break, i.e. the wire narrows at one particular point of the wire. Explain, correctly using terms like *force, stress* and *strain*, why the wire must inevitably break where the neck first forms, and why the tensile force must decrease as it does so.

Sketch stress-strain and force-extension graphs describing this behaviour, and explain why these graphs differ in shape.

5-16 The table gives the corresponding values of load and extension when masses were hung on a wire of length 2.0 m and diameter 0.40 m.

mass/kg	0	0.20	0.40	0.60	0.80	1.00	1.10	1.12
extension/mm	0	1.0	2.1	3.1	4.2	5.4	7.3	9.0

Plot a graph of load (on the *y*-axis) against extension (on the *x*-axis).
(a) Hence calculate the Young modulus of the material.
(b) Estimate the probable final extension of the wire when the load of 1.12 kg is removed.
(c) Estimate the energy stored in the wire while it carries the load of 1.12 kg.
(d) Estimate how much of this energy can be recovered.

5-17 A successfully-designed road bridge will be *strong, stiff* and *elastic*. Explain why it needs to have each of these properties, making clear the meaning of each of the words in italics.

5-18 Examine some pieces of plywood to see how the grain lies in the layers. Explain why it is put together in this way. (The Young modulus of all woods is about ten times greater along the grain than across the grain.)

5-19 In many practical situations the bending of a piece of material (e.g. a concrete beam) produces tension in the material. Draw a diagram of a concrete beam supported only at its ends, and mark on it the parts of the beam which have tensile stress in them, and the parts which have compressive stress in them.

The ultimate tensile stress of concrete is about ten times less than the ultimate compressive stress: where would you place steel reinforcing bars in the concrete beam in your diagram?

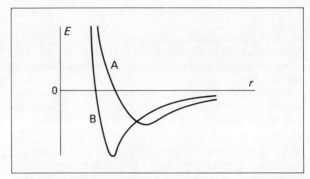

A molecular view

5-20 Explain why (a) the stiffness of rubber is low when the applied stress is low (b) rubber becomes stiffer as the stress on it increases (c) cooling rubber makes it stiffer.

5-21 Draw a graph to show how the repelling and attracting forces which two atoms exert on each other vary with their separation. What features of the graph you have drawn account for (a) the elasticity of solid materials (b) the fact that for small stresses solid materials obey Hooke's law both in tension and compression?

5-22 Explain why (a) glass-reinforced fibre is used for the bodywork of some cars (but not all) (b) glass (and not Perspex) is used for laboratory beakers (c) a plastic material (PVC) is used for rainwater pipes and gutters (d) a plastic material (polythene) is used for milk bottle crates (e) steel (coated with tin) is used for beer cans, except for the top, which is made of aluminium.

5-23 Suppose for simplicity that the copper atoms in a wire are arranged in a simple cubical lattice, the distance between rows being d.
(a) If $L = 6.02 \times 10^{23} \, \text{mol}^{-1}$, and for copper $A_r = 63.5$ and density $= 8930 \, \text{kg m}^{-3}$, calculate the number of atoms of copper per cubic metre, and hence find d.
(b) If the wire has a cross-sectional area of $1.0 \, \text{mm}^2$, how many atoms are there in any cross-section?
(c) If the wire is pulled with a force of 20 N, what is the stress in the wire?
(d) If the Young modulus of copper is 130 GPa, what is the strain in the wire?
(e) By what distance does one atom move apart from a neighbouring atom?
(f) What force does each atom exert on its neighbours?
(g) What is the 'force constant' k (i.e. the force/extension ratio) for a pair of atoms?

5-24 The figure shows how the potential energy E of a pair of atoms in a crystalline solid varies with their separation r for two elements A and B. Describe how the following physical properties differ in the solid form of the two elements, assuming that they both have the same crystal structure: (a) the size of the atoms (b) the Young modulus (c) the ultimate tensile stress.

5-25 Suppose an element exists in crystalline form in which the atoms form a simple cubical arrangement, and that the atoms are separated by a distance d (i.e. their diameter is d). A particular block of this material has m atoms along one horizontal edge, n atoms along the other horizontal edge, and is l atoms high. Vertical forces P are applied at top and bottom to stretch the block.
(a) Calculate the tensile stress in the block.
(b) Calculate the strain in the material, if the change in separation of the atoms, when the stress is applied, is Δd.
(c) Write down an expression for the Young modulus E.
(d) What is the force which each atom exerts on the other?
(e) Now think of the atoms as if they were held together by springs of stiffness k: use $F = kx$ to find an expression for the additional separation Δd when the block is stretched.
(f) Use your answers to (c) and (e) to find a relationship between E, k and d.
(g) If for copper $E = 1.3 = 10^{11}$ Pa, and $d = 2.4 \times 10^{-10}$ m, calculate k for the inter-atomic 'springs' in copper.
(h) What differences are there between the stretching of a real copper wire, and the stretching visualised in the model described above?

Failure mechanisms

5-26 The figure shows a concrete beam which is used both to support a floor in a room of a house, and also to support an overhanging balcony. Draw a diagram of the beam, showing where you would put steel reinforcing bars.

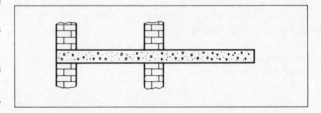

5-27 Draw a diagram to illustrate the presence of a simple (edge or line) dislocation in a crystal. With a further diagram show how applying a stress to the crystal can cause the dislocation to move. How does the presence of the dislocation make the material weaker?

5-28 Explain the following observations:
(a) Newly-drawn glass fibres are stronger than fibres which have been handled a few times.
(b) Newly-drawn glass fibres sometimes have a layer of metal (a few atoms thick) evaporated on to their surface.
(c) If a glass fibre is bent into a semi-circular arc, it may break if touched on the outside of the arc with another glass fibre, but not if touched on the inside.

5-29 Glass shatters but metals bend: explain how the molecular structure of these materials accounts for their different behaviour.

6 Fluid behaviour

Data $g = 9.81$ N kg^{-1}
atmospheric pressure = 101 kPa
density/kg m^{-3}: water, 1000; air 1.29
surface tension/N m^{-1}: water, 0.073; soap solution, 0.030

Pressure

6-1 A building brick has a mass of 2.8 kg and measures 230 mm by 110 mm by 75 mm. What pressure does it exert when stood, in turn, on each of its three faces, on a horizontal surface?

6-2 With reference to the pressure which the objects produce when they are used correctly, explain the construction of (a) skis (b) boxing gloves (c) drawing pins, (d) football boots.

6-3 Estimate the average difference of pressure there must be between the upper and lower surfaces of the wings of a jumbo jet, if the loaded plane has a mass of 350 tonnes, and wings have a total area of about 500 m^2.

6-4 In 1651 Otto von Guericke demonstrated that two teams, each of 16 horses, were needed to pull apart two hemispheres from which the air had been removed. If the hemispheres had a diameter of about 0.30 m, and we assume that with the primitive pumping apparatus of that time only 98% of the air could be removed, what was the least force which each of the horses must have exerted on the ground? (The pressure force should be calculated by multiplying the pressure by the cross-sectional area of the hemispheres, not the hemispherical area: why?

If von Guericke had later used larger hemispheres, of twice the cross-sectional area, and had still had only 32 horses, how could he have pulled the larger hemispheres apart?

6-5 If air were not easily compressible, the atmosphere would consist of a uniformly-dense (1.29 kg m^{-3}) layer of air which ended abruptly at a certain height above the Earth's surface. What height would this be?

6-6 A U-tube manometer contains oil of density 780 kg m^{-3} and has one side (A) connected to a gas supply. The other side (B) is open to the atmosphere. The difference in levels in the tubes is 230 mm. (a) On which side is the oil higher? (b) On a day when the atmospheric pressure is 101.2 kPa, what is the pressure of the gas supply?

6-7 The maximum pressure which human lungs can withstand is about 11 kPa. About how deep in water can a diver go with a snorkel tube?

6-8 The simplest form of mercury barometer consists of a vertical tube closed at its upper end. Its lower end is beneath the surface of some mercury in an open dish, and the atmospheric pressure has pushed mercury up the tube. The space above the mercury in the tube is empty, except, inevitably, for a very little mercury vapour. Explain (a) why the internal diameter of the tube does not matter (b) whether the bore of the tube need be uniform (c) whether the tube need be exactly vertical (d) whether in an accurate measurement of atmospheric pressure you would need to take account of (i) the thermal expansion of the mercury (ii) the thermal expansion of the glass (iii) the presence of water vapour in the air above the mercury in the dish (e) how you would test for the presence of air above the mercury in the tube, without using another barometer.

6-9 The figure shows two pistons P and Q which fit into tubes of cross-sectional area 4.0×10^{-4} m^2 and 1.0×10^{-2} m^2 respectively. The space between the pistons is filled with a liquid which may be assumed to be incompressible.
(a) If P is pushed with a force of 20 N, and Q is held stationary, what is the increase in pressure in the liquid?
(b) What force must be exerted on Q to keep it stationary?
(c) It is clear that with this arrangement a small push on one part of the machine can cause the machine to exert a much larger push elsewhere. Are we getting 'something for nothing'?
(d) State one advantage this type of machine will have over other machines which consist entirely of solid moving parts.

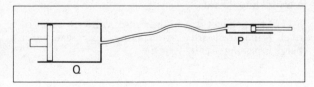

6-10 Oak has a density of about 700 kg m^{-3}. Draw a free-body diagram for a piece of oak floating in water, and explain fully why 0.7 of its total volume is submerged.

6-11 One type of hydrometer consists of a narrow stem fixed to a weighted bulb. On a particular hydrometer of this type the highest density marking (just above the bulb) was 900 kg m^{-3}: below this mark the volume was 3.0×10^{-6} m^3. What is (a) the upthrust on the hydrometer when floating in a liquid of density 900 kg m^{-3} (b) its weight (c) the upthrust on it when it floats in a liquid of density 750 kg m^{-3}? If the cross-sectional area of the stem is 4.0×10^{-6} m^2 what is the minimum length of the stem if it is to float in this liquid?

6-12 The airship R101 had a volume of 1.4×10^5 m^3. Calculate the difference between the weight of the gas in it and the upthrust on it if it were filled with (a) hydrogen (density 0.090 kg m^{-3}) (b) helium (density 0.18 kg m^{-3}).

6-13 What is the Archimedean upthrust on a body of volume 0.20 m^3 when it is suspended, completely immersed, in a tank of water?

Why is the upthrust exactly the same when it rests on the bottom of the tank?

6-14 If a piece of cork of volume 2.0×10^{-5} m^3 and density 250 kg m^{-3} is kept submerged below the surface of some water in a tank by means of a thread of negligible mass (a) what is the tension in the thread?

What would be the tension in the thread if the tank (b) were falling freely (c) were on the surface of the Moon ($g = 1.6$ N kg^{-1})?

6-15 A bathyscaphe consists essentially of a spherical steel vessel of external diameter 2.4 m, with walls 0.15 m thick. It is designed to be lowered into deep oceans. If the density of steel is 8000 kg m^{-3}, estimate (a) its weight (b) the upthrust on it, when immersed in fresh water.

To prevent it sinking it is attached to a vessel containing a liquid less dense than water (e.g. petrol, for which $\rho = 800$ kg m^{-3}). Why is air not used?
(c) Calculate the minimum volume which the vessel must have.

6-16 A stone is supported by a spring balance, and is gently lowered into a beaker of water resting on a lever balance until it rests on the bottom of the beaker, completely immersed. Describe what happens to the readings on the spring balance and the lever balance during this process.

The experiment is repeated with the beaker replaced by a Eureka can, i.e. a can with a spout fitted near the top of the can. Initially the water is level with the lip of the spout, so that as soon as the stone is lowered into the can, water begins to overflow clear of the lever balance. Describe what happens to the reading on the spring balance as the stone is lowered into the can.

6-17 An aqueduct is a bridge which carries a canal or a river. Is the downward force on the aqueduct greater when a barge is passing over it? Explain.

Flow

6-18 A horizontal pipe A of cross-sectional area 4.0×10^{-4} m^2 narrows into another horizontal pipe B of cross-

sectional area 1.0×10^{-4} m^2. The pipes are full of water, and the speed of the water in pipe A is 0.20 m s^{-1}.
(a) What is the speed in pipe B?
(b) What is the rate of flow of volume of water in (i) pipe A (ii) pipe B?
(c) Explain how the water in pipe B has been able to be accelerated.

6-19 The speed of water flowing in a pipe can be measured with a Venturi meter; a constriction is made in the pipe, and a manometer used to measure the difference in pressure between the fluid in the normal, and the fluid in the narrow, part of the pipe. When a pipe of diameter 0.20 m, carrying water, had a constriction of diameter 0.14 m inserted in it, a water manometer recorded a difference in levels of 100 mm.
(a) What was the difference in pressure in the two parts of the pipe?
(b) How many times greater was the speed in the constriction than the speed v in the main pipe?
(c) Use Bernoulli's equation to write down an equation from which you can determine v, and calculate v.
(d) What was the volume rate of flow of water in the pipe?

6-20 (a) What may happen if you drive too close to a high-sided long vehicle (e.g. an articulated truck) when you overtake it on a motorway?
(b) Suppose the sea has some waves of small amplitude. What may happen if a wind blows at right-angles to these wavefronts?

Viscosity

6-21 In order to make fluid flow at a steady speed through a pipe there needs to be a difference in pressure between the ends of the pipe, even though the fluid does not accelerate. Explain this. The pressure forces do positive work on the fluid, but it does not gain kinetic energy. What form of energy *has* increased?

6-22 The figure shows some vertical manometer tubes rising from a horizontal pipe in which a viscous liquid is flowing. Copy the diagram and sketch possible positions for the levels of the liquid in the tubes.

What would happen to the levels if the end towards which the liquid was flowing was closed?

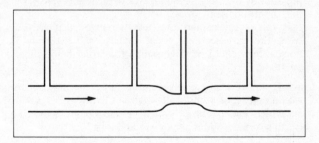

6-23 A spherical drop falling through a fluid will have a *viscous* force exerted on it by the fluid. Explain what is meant by this, and describe how the viscous force arises.

The flow of fluid past the drop will be turbulent if $\rho av > \eta$, where ρ and η are the density and viscosity of the fluid, and a and v are the radius and speed of the spherical drop. What is the fastest speed at which a raindrop can fall through air, without causing turbulent flow, if its diameter is (a) 1.0 mm (b) 0.20 mm? (Viscosity of air = 1.8×10^{-5} N s m^{-2}.)

6-24 In an experiment to measure the viscosity of some motor oil of density 820 kg m^{-3} a steel ball bearing of diameter 1.5 mm was allowed to fall through a wide vertical tube containing the oil. After reaching its terminal speed it fell a distance of 0.20 m in 28.5 s. Assuming that the flow of oil past the ball bearing is streamline, and that the viscous force can be given by $F = 6\pi a\eta v$, find the viscosity of the oil. (Density of steel = 7700 kg m^{-3}.)

6-25 Smoke rising from a lit cigarette initially rises in a steady stream, but after rising a few centimetres it breaks up into a ragged turbulent flow. Account for this change.

6-26 A small air bubble of diameter 0.20 mm rises vertically in a wide vertical tank of water. Draw a free-body diagram for the forces acting on the bubble.

If the flow of water past it is streamline, what terminal speed does it reach? (Viscosity of water = 1.0×10^{-3} N s m^{-2}.)

6-27 A horizontal capillary tube (of internal diameter 0.60 mm, and length 0.25 m) is open to the atmosphere at one end; the other end is supplied with liquid of density 900 kg m^{-3} from a tank. Liquid flows through the tube, and a volume of 2.0×10^{-5} m^3 was collected from the open end in 1.0 hour. If the level of the liquid in the tank can be assumed to be a constant height of 0.25 m above the level of the open end of the tube, find the viscosity of the liquid. Use the condition $\rho av/\eta < 1000$ to check that the flow is streamline (as it must be for Poiseuille's equation to hold).

Dimensions

6-28 Assume that the period T of a simple pendulum depends on the mass m of the bob, the length l of the thread and the gravitational field strength g. Use dimensional analysis to find how T depends on m, l and g.

6-29 Assume that the period T of a mass m oscillating vertically on a light spring of stiffness k depends on m, k and the gravitational field strength g. Use dimensional analysis to find how T depends on m, k and g.

6-30 Which of the following equations *could* be correct? b, h and s represent distances, t a time, v and V speeds, M and m masses, and g is the free-fall acceleration.
(a) $s = V^2/2g$
(b) $v(V\cos\theta - 1) = gh$
(c) $b = 4\sqrt{s}(s\cos\theta - h)$
(d) $s = h^2g/2v^2$
(e) $\dfrac{2v^2}{g}\dfrac{M + \sin\theta}{M + m}$
(f) $\dfrac{2V(M + m\sin^2\theta)}{(m + M)g\sin\theta}$.

Surface energy

6-31 How much energy is needed to increase the radius of a soap bubble from 5.0 mm to 20 mm?

6-32 A spray is used to break up 2.0×10^{-6} m^3 of water into 1.0×10^6 drops, which may be assumed to be spherical. If the initial surface potential energy of the water was 4.0×10^{-5} J, by what factor has the s.p.e. been increased?

6-33 A rectangular wire frame which measures 100 mm × 50 mm is placed in a horizontal plane and has a soap film formed in it. A wire of mass 40 mg is placed across the middle of the frame, parallel to the shorter sides of the frame, so that the soap film pulls equally on both sides of the wire. When the film on one side of the wire is broken, the wire accelerates. What is (a) the initial acceleration of the wire (b) the acceleration of the wire after it has moved 10 mm?

6-34 A vertical capillary tube has its lower end in a liquid whose density is 925 kg m^{-3} and whose surface tension is 5.5×10^{-2} N m^{-1}. What is the height of rise when the angle of contact is 0° and the internal diameter of the capillary tube is (a) 1.0 mm (b) 0.50 mm?

If on another occasion the inside of the tube was contaminated and the angle of contact was 20°, what would have been the height of rise in the tube of diameter 0.50 mm?

6-35 What is the excess pressure inside (a) an air bubble of diameter 1.0 mm, in water (b) a drop of mercury of diameter 1.0 mm, in air (γ for mercury = 0.47 N m^{-1}) (c) a soap bubble of diameter 10 mm?

6-36 Calculate the values of the excess pressure Δp inside a soap bubble for values of its radius r/mm = 2.5, 5.0, 10, 20, 30, and plot a graph of Δp against r.

Two joined soap bubbles have a common surface of spherical shape. The radii of curvature of the remaining parts of the bubbles are 6.0 mm and 15 mm. Deduce from your graph the excess pressure inside each of the bubbles. What is the difference in pressure across their common surface? Use your graph to find the radius of curvature of this surface, and draw a diagram to show which side is concave.

6-37 The figure shows a method of measuring the capillary depression of mercury. Explain how the depression h is related to γ, the surface tension of the mercury.

The density of mercury is 13 600 kg m^{-3}, and its angle of contact with glass is 140°. When the internal diameter of the narrow tube is 0.60 mm the value of h is 18 mm; find the surface tension of mercury.

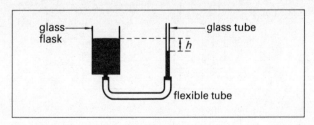

6-38 A vertical capillary tube of internal diameter 0.60 mm has its lower end 100 mm below the surface of water: the atmospheric pressure above the water surface is 100.0 kPa. The surface tension of the water is 7.5×10^{-2} N m^{-1} and the angle of contact is 0°. Find the height h of rise of the water (above the water surface) when the pressure p of the air inside the capillary tube is given by p/kPa = 99.6, 100.2, 100.8. Plot a graph of h against p.
(a) At what value of p is h zero?
(b) Up to what value of p can the straight-line graph be continued?
(c) When p = 102.0 kPa an air bubble has formed at the bottom of the tube. Calculate its radius, and draw a diagram of it.

6-39 A straight tube with a tap in the middle has a soap bubble formed at each end of the tube; the bubbles have radii of 10 mm and 15 mm respectively. Describe what happens when the tap is slowly opened so that the air in the bubbles is connected. Draw a diagram to show the final result.

6-40 The thickness of the soap film forming a soap bubble decreases with time as water evaporates from the film. Yet the size of the bubble does not change. What does this tell us about the dependence of the tension in a soap film on the thickness of the film?

6-41 A drop (of radius a) of liquid of density ρ and surface tension γ will oscillate with period T if disturbed when it is in a place where there is no gravitational field. Consider the dimensions of the quantities to find how T depends on a, ρ and γ.

6-42 Use the dimensions of the various quantities to check whether the following are possible equations for the speed c of waves on a liquid surface: (a) $c = \sqrt{(g\lambda/2\pi)}$ (b) $c = \sqrt{(2\pi\gamma/\rho\lambda)}$, where g is the gravitational field strength, λ the wavelength of the waves, ρ the density of the liquid and γ the surface tension of the liquid.

7 Electric circuits

Data g = 9.81 N kg^{-1}
e = 1.60×10^{-19} C
unified atomic mass constant m_u = 1.66×10^{-27} kg
Avogadro constant L = 6.02×10^{23} mol^{-1}

Electric current

7-1 A piece of insulated wire (whose ends you cannot reach) is fixed to the bench top. It is said to be connected in a circuit through which a current is being driven by a car battery. What simple tests could you make to confirm this if you do not want to damage the insulation of the wire? Explain how you would find the direction of the current.

7-2 Electric power is supplied to a house by a pair of overhead cables a few centimetres apart. Describe two kinds of forces with which one cable acts on the other. Would these forces tend to pull the cables together or push them apart?

7-3 In the circuit in figure (a) the current in rheostat B is twice that in rheostat C. What are these two currents?

7-4 In the circuit in figure (b) the three bulbs are identical and achieve full brightness with a current of 0.20 A. Bulb B is observed to be at full brightness. Which (if any) of the other two bulbs will be at full brightness, and what is the current in each of them?

7-5 What is the average electric current in a wire when a charge of 150 C passes in 30 s?

7-6 The current taken by a small torch bulb is 0.25 A. What total charge passes a point in the bulb circuit in 12 minutes? How many electrons pass the point in this time?

7-7 A new electric cell is joined in series with a bulb and an ammeter. The initial current is 0.30 A. At subsequent intervals of one hour the readings of the ammeter are: 0.27 A, 0.27 A, 0.26 A, 0.25 A, 0.23 A, 0.19 A, 0.09 A, 0.03 A, and at 9 hours the ammeter reading has become very small. Plot a graph of current (on the y-axis) against time (on the x-axis).
(a) Estimate (by inspection) the average current during the first 5 hours.
(b) What is the total electric charge that flows through the bulb in this time?
(c) What does the area under the graph represent?
(d) What is the total electric charge that flows through the cell in the 9 hours?

7-8 In a gas discharge tube containing hydrogen the electric current is carried partly by hydrogen ions and partly by free electrons. A milliammeter in series with such a tube indicates a current of 1.5 mA. If the rate of passage of electrons past a particular point in the tube is 6.0×10^{15} s^{-1}, find the number of hydrogen ions passing the same point per second.

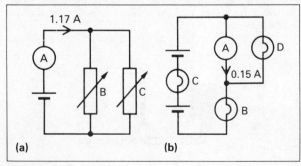

(a) (b)

7-9 The electron beam in a television picture tube travels a total distance of 0.50 m in the evacuated space of the tube. If the speed of the electrons is 8.0×10^7 m s^{-1} and the beam current is 2.0 mA, calculate the number of electrons in the beam at any one instant.

7-10 In a thunder-storm the rain drops are of average volume 3.0×10^{-8} m^3 and on average each carry a charge of 4.0×10^{-15} C. It is observed that the drops fall into a plastic bucket at an average rate of 12 per second. If the storm lasts 10 minutes (a) what is the volume of water collected (b) what is the total charge collected (c) what is the electric current into the bucket?

Currents in solids

7-11 It is usually reckoned that to avoid overheating in the copper wire of a coil the maximum allowable current per unit area of cross-section of the wire is 1.2×10^7 A m^{-2}. If there are 1.0×10^{29} free electrons per m^3 of copper, calculate the mean drift speed of the electrons when the current reaches this value.

7-12 Two copper wires of diameter 2.00 mm and 0.50 mm are joined end to end. What is the ratio of the average drift speeds of the electrons in the wires when a steady current flows in them?

7-13 The average drift speed of the electrons in a slice of n-type germanium of cross-sectional area 2.0×10^{-6} m^2. If 40 m s^{-1} when a current of 12 mA is flowing. Calculate the number of free electrons per unit volume of the material.

Currents in liquids

7-14 Calculate the specific charges of the ions $(H_3O)^+$ and $(SO_4)^{2-}$ to be found in a solution of sulphuric acid in water, given the following relative atomic masses: hydrogen, 1.008; oxygen, 16.00; sulphur, 31.97.

7-15 If the zinc ion carries a double electronic charge, how many ions of zinc are required to transport 8.0 C of electric charge in electrolysis? If this process of transport takes 20 s, what is the current?

7-16 An electric current of 10.0 A passes for 1000 s through an electrolytic tank, and is found to deposit 11.2 g of silver on one of the electrodes. What is the specific charge (i.e. the charge per unit mass) of the silver ion? If the relative atomic mass of silver is 108, what is the charge carried by a silver ion?

7-17 A steady current is passed for 50 minutes through a solution of copper sulphate, and 5.5 g of copper are deposited on the cathode. Find the value of the current. (Relative atomic mass of copper = 63.6; the copper ion carries a double electronic charge.)

Electrical energy

7-18 What is the potential difference across a piece of equipment in which 15 J of electrical energy are converted by the passage of 5.0 C of electric charge?

7-19 A lead-acid electric cell of e.m.f. 2.0 V can drive a current of 0.50 A round a circuit for 10 hours. How much chemical energy is converted to electrical energy in this time? How long would you expect the same cell to maintain a current of 0.20 A?

7-20 The e.m.f. of a small cell is 1.5 V. How much chemical energy does it convert to electrical energy when it drives a total charge of 160 C round a circuit? What energy would be converted if the same charge was driven by a battery of three such cells joined (a) in series (b) in parallel?

7-21 A potential difference of 12 V is maintained between the ends of a wire. What electric charge must flow through it to supply 960 J of electrical energy? If it takes one minute for this charge to pass, what is the current?

7-22 It is estimated that the average electric charge transported in a lightning flash is 30 C. If the energy converted in the flash is 2.4×10^{10} J, what is the potential difference involved?

In a typical thunderstorm lightning flashes strike the ground at intervals of about 3 minutes. Over the whole surface of the Earth the total current carried in this way between the atmosphere and the ground averages 1800 A. Estimate the average number of thunderstorms taking place at any one instant over the whole Earth.

7-23 One kilogram of copper is to be purified by electrolysis, the potential difference across the electrolytic tank being 8.0 V. Calculate (a) the total electric charge that must be passed (b) the energy used; in what form does this energy appear? (The relative atomic mass of copper = 63.6; the copper ion carries a double electronic charge.)

7-24 A 2.5 V bulb and a resistor are connected to a 6.0 V supply as shown in figure (a). If the bulb is at its normal

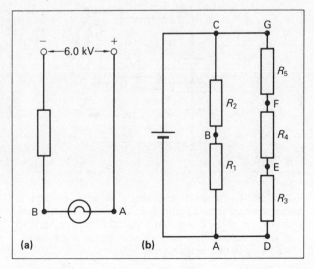

(a)   (b)

brightness, what are the potentials at A and B (a) if the negative terminal of the supply is earthed (b) if the positive terminal is earthed?

What is the potential difference across the resistor, and how much electrical energy is converted to internal energy in it when an electric charge of 200 C flows in the circuit?

7-25 Five resistors are connected as in figure (b) above to a cell. The potential differences across R_1 and R_2 are 0.60 V and 0.75 V respectively, and the resistors R_3, R_4 and R_5 are identical. Taking the zero of potential at point A, what is the potential (a) of G (b) of E? What is the potential difference (c) between B and E (d) between B and F (e) between the terminals of the cell?

7-26 In the previous question the currents in R_2 and R_5 are known to be 0.20 A and 0.30 A respectively. Write down the currents in the other three resistors and in the cell.

If these currents pass for 20 s, (a) what is the total electric charge that flows in each part of the circuit (b) what is the electrical energy converted to internal energy in each resistor.

If the e.m.f. of the cell is 1.50 V, what is (c) the chemical energy converted to electrical energy in 20 s (d) the drop in p.d. between the terminals of the cell because of the current in it (e) the electrical energy converted to internal energy in the cell?

Finally, add up all the quantities of internal energy produced at different points in the circuit, and check that this is equal to the chemical energy converted to electrical energy in the cell.

7-27 The three cells in the figure are of e.m.f. 1.2 V and are of a type across which the p.d. varies negligibly with the current in them. The three resistors are identical.

Copy the circuit diagram and mark in the potentials at each of the points A, B, C, D, E, F. Now mark in arrows to show the directions of the current in the wires leading into and out of each component. Is there any component in which there is no current?

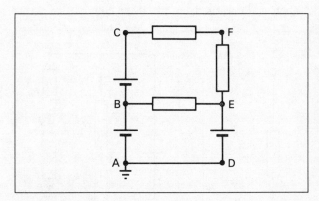

7-28 Refer to the previous question. Write down the p.d.s across each of the six components. Then add up the p.d.s round each of the loops ABEDA, BCFEB and ACFDA, and show that in each case the sum of the p.d.s round a closed loop in the circuit is zero.

7-29 What general principles can you state about (i) the currents in (ii) the potential differences across a set of electrical components in the following cases:
(a) the components are in series
(b) the components are in parallel
(c) the components are joined in a closed loop (which may however merely be *part* of a more elaborate circuit).

7-30 Two identical cells are connected to two identical resistors, as shown in figure (a); the junction C of the two cells is earthed. The p.d.s across these cells do not vary significantly with the current in them, and their e.m.f. is 1.2 V. Write down the potentials at A, B and D. In what way would the behaviour of the circuit be affected by making a connection between C and D, as shown by the dotted line?

One of the cells is now reversed, as shown in figure (b). What now are the potentials at A, B and D? [Hint: Think what the potential difference across the chain of resistors is, and therefore what the current in them must be.] What effect is now produced by making the connection between C and D?

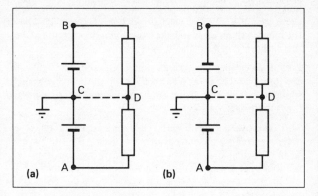

(a)   **(b)**

7-31 In the previous question, suppose that the connections are made between C and D in both cases. The current in each cell in figure (a) is known to be 0.40 A. Write down for both circuits the current in each resistor and in the link CD. Draw circuit diagrams and mark in on them the directions of these currents.

7-32 Calculate the potential gradient along the following conducting paths:
(a) a wire of length 4.0 m with a potential difference of 2.0 V applied between its ends
(b) the electrolyte between an anode and a cathode 20 mm apart between which a potential difference of 0.60 V is maintained
(c) the ionised air of a lightning flash 1.0 km long when the p.d. between cloud and ground is 20 MV
(d) the mercury vapour in a fluorescent light 1.20 m long when the p.d. across the tube is 240 V.

Electric cells

7-33 A certain type of cell has electrodes of zinc and copper; zinc ions go into solution from the first electrode, and copper ions are deposited on the second one. Which electrode is the positive terminal?

When this cell is driving a steady current in a circuit, the zinc electrode decreases in mass by 3.2 g in two hours. Calculate the increase in mass of the copper electrode in this time. (Relative atomic masses: zinc 65.4, copper 63.6. Both sorts of ion carry a double electronic charge.)

7-34 You are provided with 4 identical dry cells each of e.m.f. 1.5 V. These may be arranged to form batteries in the five different ways shown. Write down the e.m.f. in each case. The maker's instructions say that these cells should not be used with currents in excess of 0.2 A; but you are going to need to obtain a current of 0.3 A. With which of the arrangements would this be possible without damaging any of the cells?

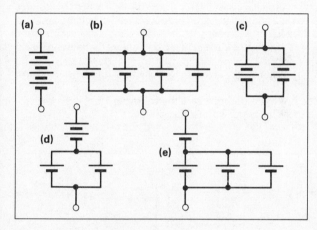

7-35 How many nicad cells each of e.m.f. 1.2 V are needed to provide a 6.0 V battery for a calculator? When the calculator is switched on, the p.d. across the battery is found to fall slightly below 6.0 V. How do you account for this in terms of the energy conversions taking place in the apparatus?

In fitting these cells in his calculator a man makes a mistake and inserts one of them the wrong way round. What will now be the e.m.f. of his battery? What energy conversions are now taking place when the calculator is switched on?

7-36 A gardener plans to make a distant-reading thermometer using thermocouples and a millivoltmeter in order to enable him to read the temperature of his greenhouse from his living room. A copper-constantan thermocouple gives an e.m.f. per unit temperature difference of its junctions of 40 μ V K^{-1}. The gardener decides to bury the cold junctions 2 m underground where the temperature remains close to 5°C throughout the year; and the temperature range of his thermometer is to be 5°C to 35°C. The millivoltmeter available gives full-scale deflection for a p.d. of 60 mV.

How many thermocouples does he need? Draw a diagram to show how the copper and constantan wires and the millivoltmeter should be connected.

Explain the energy conversions that are taking place in this arrangement.

Electrical power

7-37 What is the power consumption of a piece of equipment that converts 300 J of electrical energy to other forms in 20 s?

7-38 An electric heater takes a current of 2.5 A from a 12 V supply. What power is being supplied to it? How much internal energy is produced in the heater in 2.0 minutes?

7-39 A steady current is maintained in a small heater for 180 s, and it is estimated that the total electrical energy transformed to internal energy is 2.5 kJ. An ammeter in series with the heater gives the current as 2.4 A. What is the p.d. across the heater?

Suggest how you might calibrate a voltmeter by this kind of method. What uncertainty would you expect in the result?

7-40 How many energy is converted by a 60 W electric bulb in 5.0 minutes? What current does it take from the 240 V mains for which it is designed?

7-41 A power station generates 7.2×10^{12} J of electrical energy in a 10 hour period. What is the average power generated? If the overall efficiency of the generating process is 40%, how much internal energy must be carried away from the power station to the surroundings per hour?

7-42 A total charge of 5.0 C passes through a piece of equipment during a pulse of current lasting 10.0 ms. What is the average current during the pulse?

If the energy supplied during the pulse is 600 J, (a) estimate the average p.d. applied to the piece of equipment during the pulse (b) calculate the average power supplied.

7-43 An electric motor is being used to raise gravel from the bed of a quarry at a rate of 500 kg per minute; the gravel is lifted a total height of 100 m. If the efficiency of the motor is 70%, what current does it take from 400 V mains?

7-44 The potential difference V between the terminals of the a.c. mains supply can be represented by the equation
$$V = V_0 \sin 2\pi ft$$
Explain the meanings of the symbols V_0, f and t. What is meant by the period T of the alternating supply, and what is its value for the 50 Hz mains?

With a 50 Hz mains supply how long does it take (a) for the p.d. to grow from zero to its maximum value (b) for the p.d. to fall from its maximum value to a value half as much?

7-45 A spinning wheel has 8 spokes. What is the smallest speed at which it can appear stationary when viewed under a fluorescent light operated on the 50 Hz mains supply?

8 Electrical resistance

Data $e = 1.60 \times 10^{-19}$ C
e.m.f. of standard cell = 1.018 V
resistivities/10^{-8} Ω m:
copper 1.7, aluminium 3.2, steel 14, nichrome 130, carbon 4000

Ohm's law

8-1 There is a current of 0.20 A in a wire when the potential difference applied across it is 5.0 V. What is its resistance?

8-2 The opposite faces of a sheet of polythene are covered with metal foil. It is then found that a potential difference of 12 V between the two layers of foil causes a current of 1.4×10^{-10} A to pass through the sheet of polythene. What is its resistance?

8-3 What is the conductance of (a) the wire in question **8-1** (b) the polythene sheet in question **8-2**?

8-4 What potential difference must be applied to a resistance of 10 MΩ to drive a current of 5.0 μA through it?

8-5 The current I in a carbon composition resistor is measured for various values of the applied p.d. V, and the values shown below are obtained:

V/V	50	100	200	300	400
I/mA	0.60	1.15	2.20	3.15	4.04

What is the resistance of the resistor (a) for an applied p.d. of 400 V (b) for a very small applied p.d.?

Estimate the maximum value of the applied p.d. if the resistance is to be constant to within 10%.

8-6 A voltmeter is joined as shown in figure (a) in series with a 20 kΩ resistor across a 6.0 V battery (of negligible resistance). The voltmeter reads 3.0 V.
(a) What is the resistance of the voltmeter?
(b) What are the p.d. across the resistor and the current in it?

A second *identical* voltmeter is now joined across the resistor, as shown by the dotted lines in figure (a), and this voltmeter reads 2.0 V.
(c) What would you expect the first voltmeter now to read?
(d) What is now the current in each voltmeter?
(e) What is the current in the battery?

8-7 Two resistors R_1 and R_2 both of resistance 10 kΩ are joined in series, as shown in figure (b), across a battery of e.m.f. 6.0 V (and of negligible resistance). What are the p.d.'s across the two resistors and the currents in them?

A voltmeter is now joined across R_1, as indicated by the dotted lines, and its reading is 2.5 V. What are now the p.d.'s across the two resistors and the currents in them?

By considering the currents into and out of the junction C calculate the current in the voltmeter. Finally, calculate the resistance of the voltmeter.

8-8 The total resistance of the potentiometer between A and B in the figure is 10.0 kΩ. With the slider at the end B the voltmeter reads 2.00 V. It reads 1.00 V when the position of the slider is such that the resistance between A and C is 7.00 kΩ.
(a) What is the current in the potentiometer between C and A?
(b) What is the current in the potentiometer between B and C?
(c) What is the current in the voltmeter?
(d) What is the voltmeter resistance?

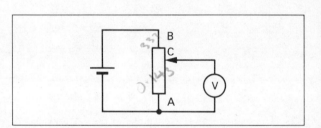

8-9 A 6 V bulb is connected in series with a resistance R, an ammeter (of negligible resistance) and a battery of e.m.f. 6.0 V (and of negligible resistance). For various values of R the following values of the current I are obtained:

R/Ω	0	2.0	5.0	10.0	20.0	50.0
I/A	0.40	0.38	0.34	0.29	0.22	0.11

Calculate and tabulate the values of the p.d. V across the bulb for each of the values of current I above, and plot a graph of I against V (i.e. the characteristic of the bulb).
(a) What is the current I for an applied p.d. V of 3.5 V?
(b) What is the resistance of the bulb (i) at 3.5 V, (ii) at 6.0 V?

Draw on the same graph the characteristic of a coil of wire of constant resistance 10 Ω.
(c) At what value of V is the resistance of the bulb 10 Ω?
(d) In the original circuit, with $R = 10$ Ω, what p.d. would the battery need to produce to give a current of 0.35 A?

8-10 The table shows the values of current I and p.d. V for a thermistor. Plot the characteristic (I against V).

V/V	0.10	0.20	0.30	0.40	0.50	0.60
I/A	0.02	0.05	0.09	0.15	0.25	0.80

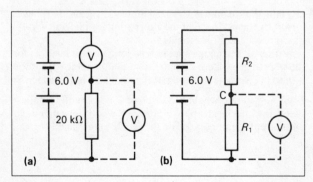

(a)   (b)

What is its resistance (a) for small values of p.d. (b) for a p.d. of 0.50 V (c) for a p.d. of 0.60 V? Calculate also the power converted in the thermistor at each of the p.d.s 0.50 V and 0.60 V.

The changes in resistance of the thermistor are caused by the changes in its temperature at various values of the power converted in it; when a p.d. is applied it takes several seconds for the temperature, and therefore the current, to reach a steady value. What will be observed if this thermistor is joined in series with the bulb in the previous question across a 6.0 V supply? Estimate the final steady current.

8-11 The current in a 10 Ω resistor must not exceed 0.40 A. What is the maximum potential difference that may be applied to it? What is then the rate of conversion of energy in it?

8-12 A potential difference of 5.0 kV is applied to a resistance of 20 MΩ. What is the current? What is the power supplied?

8-13 What is the resistance at full brightness of (a) a 240 V 60 W bulb (b) a 240 V 150 W bulb (c) a 12 V 60 W bulb?
(d) How many 12 V 60 W bulbs can be connected in series to work at full brightness across a 240 V supply?
(e) What will happen if a 240 V 60 W bulb and a 12 V 60 W bulb are joined in series across a 240 V supply?

8-14 A factory (represented by the resistor R in the figure) is supplied with electrical power from a generator through two long cables BC and AD each of resistance 3.0 Ω. The generator (connected at A and B) delivers 200 kW of electrical power at a potential difference of 6.0 kV.
(a) Calculate the current in the circuit.
(b) If A is earthed, calculate the potentials at B, C and D.
(c) What is the potential difference at which power is delivered to the factory?
(d) Calculate the rate at which electrical energy is converted to internal energy in the pair of cables.
(e) What percentage of the power is lost as internal energy in the pair of cables?
(f) If a generator delivering the same power at 2.0 kV was used instead, calculate the percentage power loss in the cables in this case. Comment on the result.

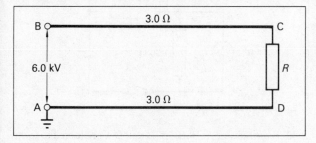

Combinations of resistors

8-15 Resistances of 2.0 Ω and 3.0 Ω are joined (a) in series (b) in parallel. What is the combined resistance in each case?

8-16 In the two arrangements of resistors shown in the figure, is the combined resistance (a) about 100 Ω (b) between 1 Ω and 100 Ω (c) less than 1 Ω? State a general rule for the order of magnitude of the combined resistance of resistors in parallel.

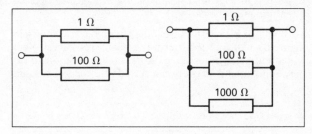

8-17 How many 240 Ω resistors must be joined in parallel to give a combined resistance of 40 Ω?

8-18 Calculate the combined resistance of each of the arrangements of standard resistances shown in the figure.

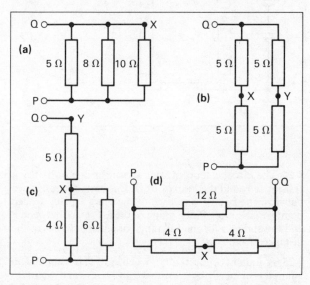

8-19 Each of the combinations of resistors in the previous question is joined across a d.c. supply, so that P is earthed and Q is at a potential of 6.0 V. Write down the potential at X in each case.

In (b) and (c) what are the potential differences between X and Y?

8-20 A nominal 2 Ω resistor is found on test to have an actual resistance of 2.040 Ω. What length of constantan wire of resistance per unit length 8.50 Ω m^{-1} should be joined in parallel with the resistor so that the combined resistance is 2.000 Ω?

8-21 Resistances of 30 kΩ and 10 kΩ are joined in series across a 12 V supply. What is the p.d. across the 10 kΩ resistor? If a voltmeter of resistance 15 kΩ is joined across the 10 kΩ resistor, what will it read? [You are advised to draw a circuit diagram before starting on the calculation.]

8-22 A battery of negligible resistance is joined to two resistances of 8.00 kΩ and 5.00 kΩ in series. A voltmeter of resistance 20.0 kΩ reads 0.800 V when joined across the 5 kΩ resistor. Calculate the e.m.f. of the battery.

What would the voltmeter read if joined across the 8 kΩ resistor? [Draw circuit diagrams of both arrangements before attempting the calculation.]

8-23 Six 1.0 kΩ resistors are joined in a circuit, as shown in figure (a), with a 12 V battery of negligible resistance. What will a voltmeter of resistance 2.0 kΩ read when joined (a) between A and B (b) between A and C (c) between B and C?

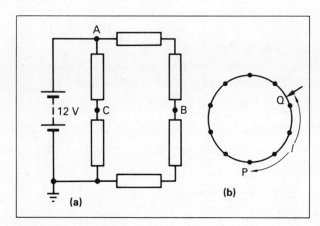

(a)   **(b)**

8-24 The circular loop of wire shown in figure (b) has a resistance round the loop of 10 Ω. Calculate the resistance between the point P and each of the other equally spaced points on the loop. Plot a graph of the resistance between P and the slider Q for the full range of possible values of the distance l from P to Q.

8-25 A student connects a 1.5 V cell, a 10 Ω resistor and an ammeter of negligible resistance in series, and wants to measure the p.d. across the resistor. He has a voltmeter of resistance 10 kΩ, but connects it in series with the other components instead of across the 10 Ω resistor.
(a) What does the voltmeter read?
(b) What is the current in the ammeter?

8-26 A meter is labelled '100 μA, 100 mV'. These are the current and p.d. which give full-scale deflection. Explain how it can be both an ammeter and a voltmeter simultaneously. If the divisions of the microampere scale are uniformly spaced, show that the millivoltmeter scale will also have uniformly spaced divisions. What is the meter resistance?

8-27 A milliammeter has a resistance of 170 Ω and gives full-scale deflection for a current of 5.00 mA. What resistance would you use and how would you connect it so that this instrument could be used (a) as an ammeter reading up to 0.5 A (b) as a voltmeter up to 5 V?

8-28 A manufacturer's catalogue lists the following moving-coil instruments.

	current at f.s.d.	resistance
(i)	50 μA	1250 Ω
(ii)	100 μA	580 Ω
(iii)	10 mA	6.00 Ω
(iv)	50 mA	0.500 Ω

Which of these instruments would you choose and why in order to construct (a) an ammeter reading up to 2 A (b) a voltmeter reading up to 1 V? In each case explain what extra resistance you would connect and draw a diagram to show how it is connected.

8-29 An ammeter of resistance of 0.040 Ω has a linear scale reading up to 0.5 A. How can it be adapted to function as an ammeter reading up to 5 A? Explain why the scale will still be linear.

8-30 The coil of a small meter has a resistance of 500 Ω; it requires a potential difference between its terminals of 150 mV to produce full-scale deflection. Explain how this instrument could be adapted to read (a) currents up to 2 A (b) p.d.s. up to 20 V.

8-31 In order to measure the calibration constant of a light-spot meter an experimenter sets up the circuit shown. The voltmeter has a resistance of 20 kΩ, and the light-spot meter a resistance of 14.0 Ω. When the switch is closed, the voltmeter reads 1.20 V and the light-spot meter deflects through 140 divisions of its scale.
(a) What is the combined resistance of the 50 kΩ resistor and voltmeter?
(b) What is the total current through the 50 kΩ resistor and voltmeter?
(c) What is the ratio of the currents in the light-spot meter and the 1.00 Ω resistor?
(d) What is the current in the light-spot meter?
(e) Hence find the calibration constant of the instrument, (i.e. the current per division of its scale).

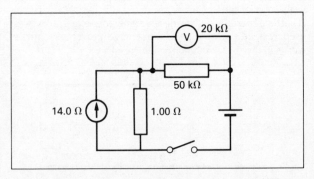

8-32 Four resistors AB, BC, CD and DA are joined to form a Wheatstone bridge (as in the figure for the next question); a meter is joined between B and D and a cell between A and C (with positive terminal at A). If AB = 4.0 Ω, BC = 5.0 Ω, CD = 6.0 Ω, and the bridge is balanced, what is the value of the resistance AD?

If this latter resistance is slightly increased, which way will current flow in the meter?

8-33 Four coils of resistance 20.0 Ω are connected as shown with a lead-acid cell of e.m.f. 2.00 V and negligible resistance. What is the p.d. between B and D? By how much does this p.d. change if the resistance CD increases by 1.0 Ω, and is B or D now at the higher potential?

The p.d. between B and D can now be brought back to its previous value by connecting another resistor in parallel with one of the other three. Which resistor should it be joined across, and what should its value be?

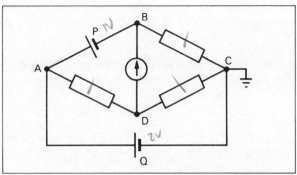

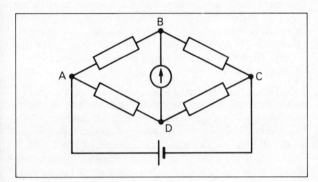

8-34 In the circuit below AB is a uniform piece of constantan wire 1.000 m long. C is a slider that may be moved to any position on the wire. It is found that there is no current in the meter when the slider C is 380 mm from A. What is the resistance *R*?

If a standard resistance of 3.00 Ω is now put in place of *R*, where should the slider C be placed for there again to be no current in the meter?

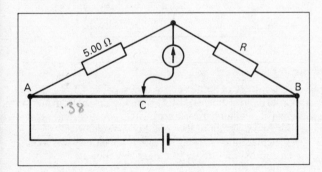

8-35 Two cells P and Q of negligible resistance are connected in the network below, together with three 1.0 kΩ resistors. The e.m.f. of P is 1.0 V and of Q 2.0 V. C is earthed.
(a) What are the potentials of A, B and D?
(b) What is the current in the meter?
(c) What are the currents in the three resistors?
(d) What are the currents in the cells P and Q?

8-36 Plot the characteristic (in the conducting direction) of a small germanium diode from the following data:

p.d./V	0.10	0.20	0.40	0.60	0.80	1.00
current/mA	0	0.1	0.8	2.3	4.3	7.2

Tabulate and plot also values of the resistance of the diode for different values of the applied p.d. At what p.d. is the resistance exactly 200 Ω?

This diode is set up in a Wheatstone bridge circuit as shown in the figure. Describe what happens as the p.d. applied between A and C is steadily increased from zero. At what value of the applied p.d. does the meter read zero? In which direction does current pass through the meter when the applied p.d. is very small?

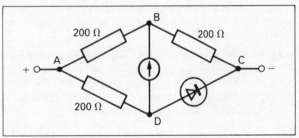

Battery resistance

8-37 The potential difference across a 1.5 V cell falls to 1.3 V when a current of 0.40 A is taken from it. What is its internal resistance? At what rate is energy converted *in the cell* (a) from chemical energy to electrical energy (b) from electrical energy to internal energy?

8-38 A dry cell is joined in a circuit in series with a switch and a resistor of resistance 2.0 Ω; and a high-resistance voltmeter is connected across the cell. With the switch open the initial reading of the voltmeter is 1.52 V. The switch is then closed and the subsequent readings *V* of the voltmeter are plotted on a graph as shown. After 6 minutes the switch is again opened and the readings are continued, as shown.

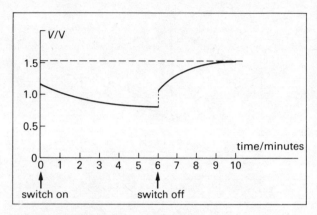

What are the e.m.f. and internal resistance of this cell (a) before any current has been taken from it (b) after it has been driving a current in this circuit for 6 minutes?

8-39 Estimate the short-circuit current that would be obtained from (a) a dry cell of e.m.f. 1.5 V and internal resistance 0.5 Ω (b) a lead-acid cell of e.m.f. 2.0 V and internal resistance 0.01 Ω.

8-40 Two cells each of e.m.f. 1.5 V and internal resistance 0.4 Ω are connected (a) in series (b) in parallel. What is the e.m.f. and internal resistance of the battery formed in each of these ways? What current would be driven by each of these batteries through a resistance of 1.6 Ω?

8-41 Refer to question **7-34** in the previous chapter. If each of the cells has an internal resistance of 0.60 Ω, what are the internal resistances of each of the battery arrangements shown in the figure?

8-42 A cell of e.m.f. 1.5 V and internal resistance 0.5 Ω is connected across a resistance of 7.0 Ω. What is the current, and what is the p.d. across the cell? How much chemical energy is converted to internal energy in 150 s (a) in the 7.0 Ω resistor (b) in the cell?

8-43 A small torch bulb is marked '2.5 V 0.3 A'. (a) What is its resistance at its normal working temperature? (b) What resistance would you join in series with it to run it from a 6 V battery? (c) When the torch bulb is run directly from a battery of e.m.f. 3.0 V, the correct p.d. of 2.5 V is produced across it. What is the internal resistance of the battery?

8-44 A tape recorder is operated by a battery of five 1.5 V cells in series, each of internal resistance 0.2 Ω. What is the combined e.m.f. and internal resistance of this battery? What is the p.d. across the battery when a current of 150 mA is being taken?

8-45 A battery of thirty 2 V lead-acid cells in series, each of internal resistance 0.050 Ω, is to be charged from an 80 V d.c. supply. The charging current is to be 2.0 A. What resistance should be connected in the circuit? Draw a circuit diagram of the arrangement. What is (a) the power taken from the supply (b) the rate of production of internal energy in the battery and in the resistor? How do you account for the difference between (a) and (b)?

8-46 You want to charge a 6 V motor cycle battery using a 12 V car battery; both batteries are of negligible internal resistance. If the charging current is to be not more than 1.0 A, what additional resistance would you require? Draw a circuit diagram to show how you would connect it up to the batteries.

If the only resistances available were an assortment of motor cycle and car light bulbs, state the operating p.d. and power of the bulb you would select for the job.

8-47 A battery of six lead-acid cells in series is to be charged by connecting it directly to a small d.c. dynamo. Initially the e.m.f. of each cell is 1.8 V, but this rises to 2.5 V when the cell is fully charged. The charging current is never to be more than 2.0 A, and this is to fall to zero when the battery is fully charged. What e.m.f. and internal resistance should the dynamo have?

8-48 The circuit shown represents the electrical system of a motor car with the engine running steadily and the lights switched on. G is a d.c. dynamo of internal resistance 1.5 Ω; the resistance R of 1.40 Ω represents the combined resistance of the lights. The lead-acid battery is of e.m.f. 12.5 V and internal resistance 0.050 Ω. If a voltmeter joined across the battery reads 12.6 V, what is (a) the current taken by the lights (b) the current in the battery (c) the e.m.f. of the dynamo?

If now the wire making connection to one terminal of the battery breaks, what does the p.d. applied across the lights become (assuming that the resistance of the lights stays the same)?

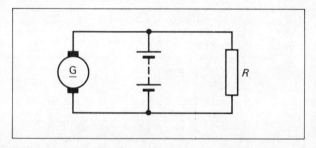

8-49 A battery of 5 nife cells in series, each of e.m.f. 1.2 V and internal resistance 0.040 Ω is to be charged from a 12 V supply (of negligible resistance).
(a) Draw a diagram of the circuit you would use.
(b) Calculate the charging current if the battery was joined *directly* across the supply (the right way round).
(c) Calculate the current if the battery was joined across the supply the *wrong* way round.
(d) Calculate the series resistance needed to limit the charging current to 5.0 A.
(e) With this resistance in the circuit calculate the p.d. between the terminals of the battery and the rate of production of internal energy in the battery.

8-50 A small sealed capsule has two external terminals. It is thought to contain resistances and either (a) a small cell or (b) a semiconductor diode. When the terminals are joined to a 2.0 V cell of negligible resistance the current is found to be 0.10 A. When the cell connections are reversed, the current is 0.40 A. Suggest what could be in the capsule in both alternatives (a) and (b).

What further tests would you perform to decide between these alternatives?

Resistivity

8-51 A uniform wire is of length 4.0 m and cross-sectional area 1.0×10^{-6} m^2. The resistivity of its material is 4.8×10^{-7} Ω m. What is its resistance?

If this wire carries a steady current of 0.50 A, what is the potential gradient along the wire?

8-52 When a potential difference of 100 V is applied between opposite faces of a slab of polythene 5.0 mm thick and 0.50 m square, a current of 2.5×10^{-12} A is found to flow through it. What is the resistivity of polythene? What is its conductivity?

8-53 The heating element of an electric toaster is designed to convert 400 W of electrical power in a 240 V circuit. It is made of nichrome ribbon 1.0 mm wide and 0.050 mm thick. Calculate the length of ribbon required. (See table of resistivities above.)

8-54 A power cable consists of six aluminium wires enclosing a central steel wire. (The steel wire gives the cable mechanical strength, but carries a negligible proportion of the current.) If each of these wires is of diameter 4.00 mm, calculate (a) the cross-sectional area of each wire (b) the resistance of 1.0 km of aluminium wire (single strand) (c) the resistance per km of the whole cable.

8-55 Calculate the resistance between opposite edges of a square film of conducting material of side l, thickness d and resistivity ρ. Show that the resistance is independent of l.

What is the resistance of such a square film of carbon 5.0×10^{-7} m thick deposited on an insulating plate?

8-56 A copper wire and a steel wire are of identical length and diameter. What is the ratio of their resistances?

These wires are connected first in series and then in parallel to a supply of variable p.d. As the p.d. is increased it is found that in the first case the iron wire starts to glow before the copper; but in the second case (in parallel) the copper glows first. How do you explain this?

8-57 The heating element of an electric fire is designed to consume 1.0 kW of electrical power on a 240 V supply with the element at a temperature of 900°C. When the element is disconnected and cooled to 0°C its resistance is found to be 50 Ω. What is the average temperature coefficient of resistance of the metal over the range from 0°C to 900°C?

8-58 Draw a circuit diagram showing how you would use an ammeter, a voltmeter and a rheostat (with a suitable supply) to investigate how the current I through a 12 V bulb varies with the applied potential difference V. The set of readings below is obtained in such an experiment.

V/V	0.2	0.5	1.0	2.0	4.0	6.0	8.0	10.0	12.0
I/A	0.10	0.17	0.22	0.26	0.32	0.37	0.42	0.46	0.50

Plot the characteristic of the bulb (i.e. a graph of I against V). Deduce from your graph the resistance of the bulb at room temperature. By what factor does the resistance increase between room temperature and the maximum temperature of the filament (at 12.0 V)? Is this in agreement with the observation that with some metals resistivity is proportional to kelvin temperature (take the maximum temperature of the filament as 2400 K)?

8-59 On the graph of the previous question draw also the characteristic of a wire of constant resistance 20 Ω.
(a) At what p.d. does the bulb have a resistance of 20 Ω?
(b) If the bulb and the wire are joined in series, what p.d. would be required to drive a 0.40 A through them?
(c) What p.d. would be required to drive a current of 0.35 A through two such bulbs in series?
(d) What would be the total current if a p.d. of 11 V was applied across two such lamps in series?
(e) What would be the total current if a p.d. of 11 V was applied across two such lamps in parallel?

8-60 The density of free electrons in copper is 1.0×10^{29} m^{-3}; (its resistivity is given in the table above). Calculate the average drift speed of the electrons when a potential gradient of 0.10 V m^{-1} is applied to the wire. What value does this give for the mobility of the electrons?

8-61 The filament of a bulb is of length l and diameter d; it is of resistivity ρ at its operating temperature. If it carries a current I, calculate (a) the rate of conversion of electrical energy to internal energy in the filament (b) the surface area of the filament (c) the rate of radiation of energy per unit area of the filament surface.

Calculate the latter quantity for a tungsten filament of diameter 0.080 mm carrying a current of 1.0 A. (The resistivity of tungsten at the operating temperature of the filament is 5.6×10^{-7} Ω m.)

Potentiometer circuits

8-62 A steady current is driven through a potentiometer wire 1.000 m long. One terminal of a standard cell is connected as shown in the figure to one end A of the wire, and the other terminal is connected through a sensitive meter to a contact maker that may be moved to any point on the wire.

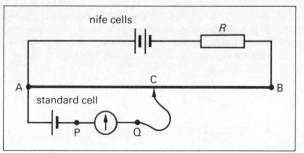

It is found that there is no current in the meter when the contact maker C is 566 mm from A. A cell of unknown e.m.f. is now substituted for the standard cell, and the balance length in this case is found to be 773 mm. What is the unknown e.m.f.?

8-63 In the previous question, taking the zero of potential at A, write down the potentials at P and Q when the potentiometer circuit is balanced with a standard cell. Explain why the resistance of the meter does not affect the balance point C of the contact maker on the wire. What advantage is there then in having a meter of low resistance in this circuit?

8-64 A standard resistor of resistance $100.0 \, \Omega$ and an unknown resistance R are joined in series and a steady current is passed through them. Using a potentiometer circuit with potentiometer wire 1.000 m long in the usual way, a balance length of 692 mm is obtained with the potentiometer circuit joined across the standard resistor; but joined across the unknown resistance R the balance length is 869 mm. Calculate R.

8-65 A steady current is passed through a standard resistance of $0.500 \, \Omega$ and an ammeter joined in series, and the current is adjusted so that the ammeter reads exactly 1.50 A. The p.d. across the standard resistance is then found to be balanced by the p.d. across 680 mm of a potentiometer wire. The wire has previously been calibrated so that the potential gradient along it is $1.00 \, \text{V m}^{-1}$. What is the error in the ammeter at this point of its scale?

8-66 When setting up a simple potentiometer circuit such as that shown in the figure a common procedure used by experimenters to test the circuit before attempting to make measurements is the following. With a protective resistor in series with the meter the contact maker C is placed first at one end A of the potentiometer wire and then at the other end B, and in each case the behaviour of the meter is noted.
(a) What should the meter do in this test?

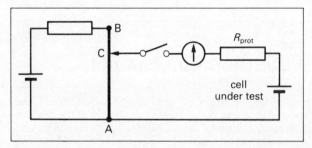

If, however, a mistake has been made in setting up the apparatus, the meter will behave differently.
(b) **Explain in each case below how the meter will behave, and which deflection (at A or at B) will be the greater.**
(i) The cell under test is connected the wrong way round.

(ii) The resistance in series with the potentiometer wire is too great.
(iii) There is a broken connection in the wires near the driver cell.
(iv) There is a broken connection in the wires near the meter.

8-67 A lead-acid cell of e.m.f. 2.0 V is joined by wires of negligible resistance across a uniform potentiometer wire of length 1.000 m and resistance $2.0 \, \Omega$. One terminal of a cell of e.m.f. 1.5 V and internal resistance $0.80 \, \Omega$ is joined to one end of the wire, and the other terminal of the cell is connected through a meter to a slider on the wire. At what position of the slider will the meter read zero? Draw a circuit diagram.

What balance lengths of the wire will be required for zero meter reading if a resistance of $0.50 \, \Omega$ is joined (a) across the 1.5 V cell (b) in series with the lead-acid cell and the potentiometer wire? Again draw circuit diagrams of each of these arrangements.

Two experimenters perform the above measurements with apparently identical apparatus. The first finds that in both (a) and (b) the balance lengths steadily increase with time. The second finds that in (b) the balance length stays steady, but that in (a) it falls slowly. Explain these results.

8-68 A thermocouple of resistance $15 \, \Omega$ is placed with one of its junctions in melting ice and the other in the steam above boiling water. Its e.m.f. measured with a potentiometer circuit is 4.0 mV. What reading would be obtained if it is connected to a millivoltmeter of resistance $45 \, \Omega$?

8-69 You are provided with a standard cell and the components with which to construct a potentiometer circuit, together with a good stock of standard resistors, etc. How would you arrange to measure the e.m.f. of a thermocouple (known to be about 4 mV)? Give suitable values for the components in your circuit, explaining why you have selected those particular values.

8-70 Using the same stock of equipment as in the previous question, explain how you would construct a potentiometer circuit to measure the p.d. produced by a high-voltage unit (known to be about 5 kV, and of internal resistance $5 \, \text{M}\Omega$). Again give suitable values for the components, and explain why you have selected those values.

8-71 A potentiometer circuit is made up of a 2.00 V cell in series with a resistance R and a uniform wire AB of length 1.000 m and resistance $2.00 \, \Omega$. A thermocouple of e.m.f. 4.00 mV is connected through a sensitive meter to A and to a slider on the wire. No current is shown by the meter when the slider is exactly at the mid-point of AB. What is the value of R?

If now the resistance R is increased by $2 \, \Omega$, how far would the slider have to be moved to obtain zero current in the meter again?

9 Heating solids and liquids

Data $g = 9.81$ N kg^{-1}
Avogadro constant $L = 6.02 \times 10^{23}$ mol^{-1}
density/kg m^{-3}: water 1000, air 1.29
linear expansivity (290 K)/10^{-5} K^{-1}: brass 1.90, iron and steel 1.2, copper 1.68
s.h.c. of water $= 4.18$ kJ K^{-1} kg^{-1}
s.l.h. of fusion of ice $= 0.334$ MJ kg^{-1}
s.l.h. of vaporisation of water $= 2.26$ MJ kg^{-1}

Temperature and internal energy

9-1 Sketch a graph of the variation of the intermolecular potential energy of two molecules with their separation, labelling any features which you think are important. Use the graph to explain how the potential energy varies for a molecule in a solid which is (a) a few degrees above absolute zero (b) a few hundred degrees above absolute zero.

9-2 Explain the advantages you think the kelvin scale of temperature has over the Celsius scale, but why most thermometers in a laboratory are marked in Celsius degrees.

9-3 *Heating* and *working* are two different ways in which energy can be transferred from one body to another. Would you describe the following energy exchanges as heating or working? State in each case the form of energy lost by the body giving the energy, and the form of energy gained by the body receiving the energy.
(a) A can of beer is taken from a warm room and put in a refrigerator.
(b) A man sandpapers a block of wood: its temperature rises.
(c) A night storage heater cools down during the day.
(d) A boy pumps air into a bicycle tyre: the pump gets hotter.
(e) A tennis ball is dropped and after several bounces comes to rest.
(f) The coffee in a cup is rotating, having just been stirred. After some time it comes to rest.

9-4 When a car's brakes are applied frictional forces do 0.20 MJ of work. Because they are hot they lose 0.080 MJ of energy to the surroundings. What are (a) ΔW (b) ΔQ (c) ΔU for this process?

9-5 What is meant by the *internal energy* of a body? A car is stopped by being braked. In what sense is the increase of energy of the molecules of the brake drums, tyres, road, etc., different from the original k.e. of the car?

9-6 A reservoir gains water from a stream, and by rain falling on it. It loses water by evaporation, and when called upon to supply water to a town. Visiting the reservoir you ask the resident engineer 'how much rain is there in the reservoir?' He thinks this is a pointless, and not very sensible, question.

This situation can be used as an analogy to help us understand the use of the term *internal energy*, and the concepts of *heating* and *working* as ways of supplying or removing energy. Explain, using this analogy, why it is pointless to ask 'how much heat is there in this body?'.

Expansion

9-7 (a) What is the increase in length of the Forth (railway) Bridge which is 521 m long and made of steel, when the temperature changes from $-5°$C to $25°$C?
(b) What is the increase in volume of 1.0×10^{-3} m^3 of water when its temperature rises from $20°$C to $100°$C? (γ for water $= 2.1 \times 10^{-4}$ K^{-1}.)

9-8 In an experiment to measure the linear expansivity of aluminium a bar of it was raised in temperature from $18.5°$C to $99.5°$C. The length of the bar was 0.50 m. One end of the bar was fixed; the other end was arranged so that a micrometer screw gauge could measure its distance from a fixed point. Before and after heating the micrometer screw gauge readings were 2.31 mm and 3.24 mm. What value for the linear expansivity of aluminium do these readings give?

9-9 A rectangular slab of copper measures 50 mm by 40 mm and has a rectangular hole which measures 40 mm by 30 mm, symmetrically placed. It is not, of course, possible to raise the temperature of the copper so much that the expansion is even visible by eye, but *if* it could be raised in temperature by 50 000 K, what would be the new length of the 50 mm side?
Draw a full-size diagram of the slab as it originally was, and superimpose a diagram of the hotter slab.

9-10 An iron ring at $0°$C has external and internal diameters of 150.00 mm and 120.00 mm respectively. If it is heated to $100°$C, what are their new values? What is the increase in area of the hole?

9-11 Rail for railway track was once laid in lengths of about 18 m (60 ft). If it was laid at $10°$C what should the gap be between adjacent lengths of rail to allow the temperature to rise to $30°$C without the rail buckling?
Track is now welded into lengths which are often more than 1 km: it is heated to a temperature of $30°$C and bolted to the sleepers while still hot. When it cools the rail is in tension. Explain how this prevents the track buckling when the air temperature rises.
Calculate (a) how much a 1.0 km length of rail would contract if it were free to do so, and the temperature fell from $30°$C to $5°$C (b) the stress needed to compress the rail by this amount (E for steel $= 200$ GPa) (c) the tensile stress in the rail at $5°$C.

9-12 What percentage increase in the spacing of copper atoms in a crystal would you expect when the copper is raised in temperature by 100 K?

9-13 When 1.50×10^{-3} m^3 of oxygen is raised in temperature from 273 K to 450 K its volume increases to 2.47×10^{-3} m^3 if its pressure remains constant. What is the cubical expansivity of oxygen?

9-14 The density of ethanol at 273 K is 789 kg m^{-3}, and its cubical expansivity is 1.12×10^{-3} K^{-1}. What is its density at 323 K?

9-15 A careful examination of the reading of a mercury thermometer will show that the reading initially falls (slightly) when it is placed in a hot liquid. Why does this happen?

9-16 The table gives the volume of 1.000 00 g of water at various Celsius temperatures θ. Calculate the density ρ of the water at each temperature, and draw a graph of ρ against θ. On the ρ-axis the scale should cover the range 1.000 00 to 1.000 30; it should *not* include the origin.

θ/°C	0.0	2.0	4.0	6.0
V/10^{-6} m^3	1.000 16	1.000 06	1.000 03	1.000 06

θ/°C	8.0	10.0
V/10^{-6} m^3	1.000 15	1.000 30

Ponds freeze at the surface only, unless the air temperature is below 0°C for a long time. Consider what happens to the temperature and density of the water in an undisturbed pond as the temperature of the air above it falls from 10°C to − 10°C and explain this phenomenon.

9-17 Explain why, if very hot water is poured into (a) a thick glass jar it may crack (b) a thin glass jar it probably will not crack (c) a thick Pyrex jar it will not crack.

9-18 Sketch a graph of the variation of the intermolecular potential energy of two molecules with their separation, and use it to explain why solids expand when their temperature rises. Label your graph A. Draw a second graph, labelled B, for a material which has a lower linear expansivity.

Heat capacity

9-19 What are the heat capacities of the following? (a) 2.0 litres of water (b) a kettle which consists of 0.25 kg of aluminium and a heating element which contains 0.20 kg of iron (c) a copper calorimeter of mass 0.10 kg (d) an expanded polystyrene cup of mass 10 g (S.h.c.s/J K^{-1} kg^{-1}: aluminium 878, copper 380, iron 440, expanded polystyrene 1400.)

9-20 An 'instant' gas water heater is capable of raising the temperature of 2.0 kg of water by 50 K each minute. What is its power?

What problems might there be in designing an instant electric water heater which is to work from the ordinary mains supply?

9-21 Make a rough calculation of the cost of using electrical energy to heat water for a bath, if about 0.3 m^3 of water have to be heated from 5°C to 35°C. Assume that the cost of 10^7 J of electrical energy is about 7p.

9-22 It is sometimes said that the cost of running an upright freezer is greater than the cost of running a chest (horizontal) freezer because each time the door of an upright freezer is opened all the cold air falls out. Consider an upright freezer of internal capacity 0.20 m^3 and discuss whether this is likely to be true. The s.h.c. of air at constant volume is about 600 J K^{-1} kg^{-1}, and the temperature inside the freezer is likely to be about − 18°C.

9-23 A 100 W immersion heater is placed in 200 g of water in a plastic cup (of negligible h.c.). How fast does the temperature rise (ignoring the heating of the surroundings)?

9-24 How long would it take a 2.0 kW heater to warm the air in a room which measures 4 m × 3 m × 2.5 m from 5°C to 20°C. The s.h.c. of air under these conditions is about 1000 J K^{-1} kg^{-1}.

In practice it takes *much* longer (perhaps an hour?) for a heater to warm a such a room. Why is this?

9-25 A car of mass 800 kg moving at 20 m s^{-1} is braked to rest 10 times. If 20% of the car's kinetic energy is given to the steel brake drums, what is their rise in temperature, if each of the four has a mass of 1.5 kg? (S.h.c. of steel = 500 J K^{-1} kg^{-1}.)

9-26 Water has a relatively high s.h.c. What effect does this have on (a) climate (b) the cost of heating water for baths and showers (c) the volume of water in a car engine's cooling system (d) the running costs of a water-filled central heating system?

9-27 1.0 kg of water at a temperature of 95°C is poured into a copper saucepan of mass 0.70 kg which is at a temperature of 20°C. What is the temperature of the water immediately after it has been poured? (S.h.c. of copper = 380 J K^{-1} kg^{-1}.)

9-28 A block of copper of mass 400 g is raised to a temperature of 450 K and lowered into 500 g of water contained in a vessel of h.c. 200 J K^{-1}; these were initially at a temperature of 290 K. If the final temperature of the water was 300 K, what value do these readings give for the s.h.c. of copper?

What precautions would you take during such an experiment?

9-29 *Without* consulting a data book, place the following in order of decreasing s.h.c.: water, lead, copper, aluminium, ethanol.

9-30 A kettle of h.c. 500 J K^{-1} contains 1.5 kg of water at 12°C. If its power is 2.5 kW, how long will it take for the kettle and water to reach 100°C, ignoring losses of energy to the surroundings?

If, in practice, having reached 100°C and after being switched off, the kettle cooled to 96°C in 50 seconds, what was (a) the rate of loss of energy at 98°C (b) the average rate of loss of energy while the temperature was rising from 12°C to 100°C (c) the loss of energy during this time?

9-31 A 30 W immersion heater is placed in an aluminium block of mass 1.0 kg, and switched on for 6.0 minutes: the initial temperature of the block and the surroundings was 12.3°C. At successive one-minute intervals, the following temperatures were recorded: 14.1°C, 15.9°C, 17.7°C,

19.5°C, 21.2°C, 22.9°C, 22.8°C, 22.7°C, 22.6°C, 22.5°C. Estimate the maximum temperature which would have been reached if the block had not been heating the surroundings, and hence calculate the s.h.c. of aluminium.

What sources of error remain, and do they suggest that the value obtained is an over-estimate or an under-estimate?

9-32 A cubical block of side 200 mm at a temperature of 45°C loses energy to the surroundings at a rate of 10 W when the surroundings are at a temperature of 15°C. What would be the rate of loss of energy for a block of side 100 mm if its temperature were 75°C.?

9-33 0.30 kg of a liquid of s.h.c. 1700 J K^{-1} kg^{-1} was placed in a container of h.c. 90 J K^{-1} and allowed to cool in surroundings whose temperature was 20°C. The table shows corresponding values of temperature and time:

t/min	0	2.0	4.0	6.0	8.0	10.0	12.0
θ/°C	95.3	87.8	81.0	74.9	69.4	64.5	60.0

Plot a graph of θ against t.
(a) Measure from your graph the rates of fall of temperature at t = 2.0 and t = 10.0 minutes
(b) What was the *excess* temperature of the liquid, above its surroundings, at these two times?
(c) Is the rate of fall of temperature proportional to the excess temperature?
(d) Explain why this is to be expected.
(e) What was the rate of loss of energy at 87.8°C?

9-34 A 30 W heater is placed in a block of aluminium. When room temperature is 17°C the final steady temperature reached by the block is 86°C. What is the rate of loss of energy by the block when its temperature is (a) 86°C (b) 40°C?

If the heat capacity of the block was 900 J K^{-1}, what would be the rate of fall of temperature of the block as soon as the heater was switched off (at 86°C)?

9-35 In an X-ray tube electrons are accelerated by a potential difference of 100 kV and strike a target: it may be assumed that all the kinetic energy of the electrons becomes internal energy in the target. Water is circulated through the target to keep its temperature constant. The current in the beam of electrons is 5.0 mA.
(a) What is the rate at which electrical potential energy is given to the electrons?
(b) What is the rate at which the circulating water gains internal energy?
(c) If the temperature rise of the water is 20 K, at what rate must it flow through the target?

9-36 An oven of heat capacity 4.0 kJ K^{-1} has a 100 W heater which is thermostatically controlled so that the heater is switched on when the temperature falls to 62°C and switched off when the temperature has risen to 64°C. It is found that the heater is switched on for periods of 95 s.
(a) How much energy is supplied by the heater in that time?
(b) How much energy is needed to raise the temperature of the oven from 62°C to 64°C?
(c) What is the average rate at which energy is being given to the surroundings while the heater is switched on?

(d) How long are the periods for which the oven is switched off?

9-37 A wide cardboard tube is fitted with corks at both ends and 0.20 kg of lead shot is placed in it. The inside length is 0.80 m. It is held vertically, so that the lead shot is at the bottom of the tube, and then quickly inverted, so that the lead is lifted 0.80 m, and then falls through that distance. After 150 such inversions the temperature of the lead is found to have risen by 6.5 K.
(a) What value does this give for the s.h.c. of lead?
(b) Would you expect this to be an overestimate or an underestimate?
(c) Why is lead shot used? Are there any alternatives?
(d) The temperature rise is small! Would it help to increase the mass of lead shot?

9-38 The figure shows an apparatus for measuring the specific heat capacity of a (solid) metal. The handle is rotated against the frictional torque of the rope wrapped round the cylinder of the metal. The work done warms the cylinder, and its temperature rise can be measured.

In a particular experiment the handle was turned 100 times, the load was 10.0 kg, and the rubber band was slack. The cylinder was made of copper; its diameter was 25.0 mm and its mass was 0.180 kg. The temperature rose from 13.15°C to 23.20°C. What value do these measurements give for the s.h.c. of copper?

If the allowance of 3.5 J K^{-1} is made for the h.c. of the thermometer, what value is now obtained for the s.h.c. of copper?

What other major source of error is there? Describe how you would allow for it.

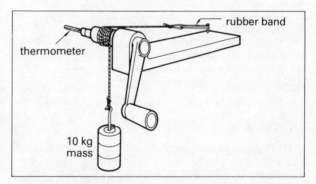

9-39 A simple method of measuring the s.h.c. of a liquid would be to place some in a metal can and use an immersion heater to raise its temperature. Why would you not expect this to give an accurate value for the s.h.c., and what improvements would you make? You may assume that the power of the heater is accurately known.

9-40 In a school laboratory continuous-flow method of measuring the s.h.c. of water the following measurements were made:

temperature of water at inlet and outlet: 12.1°C, 14.5°C
mass of water collected in 5.0 minutes: 231 g
ammeter and voltmeter readings: 2.3 A, 4.0 V

What value do these measurements give for the s.h.c. of water? This value is only within 20% of the accepted value for water, but such inaccuracy is to be expected with this simple form of the apparatus. Where would you look for sources of inaccuracy?

9-41 What are the advantages of a continuous flow method of measuring the s.h.c. of a liquid?

9-42 A current of 1.2 A is passed through a resistor which is immersed in water in a calorimeter. The mass of the water is 0.20 kg, and the h.c. of the calorimeter is 40 J K^{-1}. It may be assumed that losses of energy to the surroundings are negligible. After 10 minutes the temperature of the water has risen by 6.7 K. What is the potential difference between the ends of the resistor? Explain how this experiment could be used to calibrate a voltmeter.

9-43 4.0 g of argon were placed in a cylinder fitted with a piston. The piston was held fixed, and when a heater of power 5.0 W was used for 40 s the temperature of the argon rose from 260 K to 354 K. The piston was then released, and when the same heater was used for the same time the temperature rose from 320 K to 377 K. Calculate the two values of the s.h.c. of argon which these measurements give, and explain why they are different.

9-44 The table gives corresponding values of the s.h.c. of some elements (at 290 K) and their molar masses.

element	Mg	Al	Ca	Ti
c_p/J K^{-1} kg^{-1}	980	877	636	473
M_m/kg mol^{-1}	0.0243	0.0270	0.0401	0.0479

element	Zn	Mo	Cd	U
c_p/J K^{-1} kg^{-1}	384	301	229	117
M_m/kg mol^{-1}	0.0654	0.0959	0.112	0.238

(a) Calculate values of $1/M_m$ and plot values of c_p (on the y-axis) against $1/M_m$ (on the x-axis).
(b) Would it be justifiable to draw a straight line through the points you have plotted?
(c) Explain why the slope of such a line gives a value of $c_p M_m$.
(d) What are the units of this quantity, and what is the value?
(e) What is its significance?

9-45 The molar heat capacity of many elements in the solid phase is nearly 25 J K^{-1} mol^{-1} at room temperature. How much energy is needed to raise by 10 K (a) 2.0 mol of aluminium (b) 2.0 mol of copper (c) 1.0 mol of aluminium (d) 1.2×10^{23} atoms of aluminium (e) 1.2×10^{23} atoms of copper?

9-46 The s.h.c. of copper (at constant pressure) varies with temperature as shown in the table:

T/K	25	50	100	150	200	300	400
c_p/J K^{-1} kg^{-1}	11	99	255	323	357	386	395

Plot a graph of c_p (on the y-axis) against T (on the x-axis).
(a) What does the area between the graph and the T-axis represent?
(b) How much energy is needed to raise the temperature of 0.30 kg of copper from (i) 50 K to 150 K, (ii) 250 K to 350 K?

Latent heat

9-47 How much energy must be (a) given to 2.0 litres of water at 100°C to evaporate it, (b) taken from 0.50 kg of water at 0°C to freeze it?

9-48 How long will it take (a) a 1000 W heater to evaporate 1.0 kg of water which is already at 100°C, its n.b.p.? (b) a refrigerator to freeze 1.0 kg of water which is already at 0°C, its n.m.p., if it can remove energy at a rate of 75 W?

9-49 An immersion heater is embedded in ice in a vessel at -10°C and switched on until all the water formed has evaporated. Sketch a graph to show the variation of temperature in the vessel with time. Explain briefly in molecular terms what happens to the energy supplied by the heater at different parts of the graph.

9-50 Explain the following:
(a) sponge pudding is often served with a hot sauce, and meat is served with hot gravy
(b) it is advisable to put on extra layers of clothing *after* one has played an energetic game, when one is already hot.

9-51 How much hot water, at a temperature of 90°C, must be poured on to 0.40 kg of ice at 0°C to melt it?

9-52 In an 'expresso' coffee-making machine steam at 100°C is passed into 0.18 kg of cold coffee (s.h.c. the same as that of water) to warm it. If the initial temperature of the coffee is 14°C, what mass of steam must be supplied to raise the temperature of the water to 85°C?

9-53 Two lumps of ice at 0°C, each of mass 20 g, are added to a glass containing a mixture of alcohol and water at a temperature of 15°C. The h.c. of the glass and its contents is 600 J K^{-1}. When the system has reached equilibrium how much ice is there?

9-54 The mass of liquid nitrogen in an open beaker is found to have decreased by 46.3 g in 10 minutes. If the s.l.h. of vaporisation of nitrogen at its boiling point is 1.99×10^5 J kg^{-1}, at what rate were the surroundings heating the beaker? Why is the heat capacity of the beaker irrelevant?

9-55 A kettle of h.c. 500 J K^{-1} contains 1.3 kg of water at 100°C. With the element (2500 W) switched off it takes 15 s for the temperature to fall to 99°C. If switched on again, how long after it has again reached 100°C will it boil dry?

9-56 The figure shows apparatus which can be used for a continuous flow method of measuring the s.l.h. of vaporisation of a liquid. After waiting for conditions to become steady 47 g of liquid were collected in 40 minutes, when the ammeter and voltmeter readings were 4.11 A and 6.10 V respectively. What value do these measurements give for the s.l.h. of vaporisation of the liquid?

The experiment was then repeated using less power. The ammeter and voltmeter readings were now 3.27 A and 4.90 V, and the mass collected in 40 minutes was 28 g. Use the two sets of readings to obtain a more accurate value for the s.l.h., and calculate the rate of loss of energy from the apparatus.

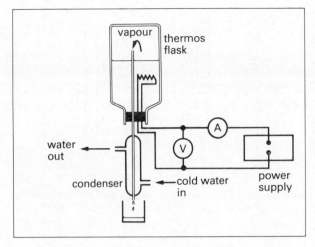

This technique makes allowance for the loss of energy from the apparatus. Why was it necessary (a) for the experiment to have been performed for the same length of time on both occasions (b) to wait until conditions became steady?

9-57 In the steel-making process the solid raw material (iron) has to be raised to a very high temperature, about 1800 K, when the iron melts. Calculate (a) the energy needed to raise 40 kg of iron from 300 K to 1800 K (b) the energy needed to melt 40 kg of iron (c) the total energy needed (d) your answer to (b) as a percentage of your answer to (c). (Average s.h.c. of iron = about 700 J K^{-1} kg^{-1} over the range 300 K to 1800 K, s.l.h. of fusion of iron = 270 kJ kg^{-1}.)

9-58 An open dish of liquid is very slightly cooler than its surroundings. Why? Your explanation should include an account of why its temperature is steady, and the factors which determine that steady temperature.

9-59 The following temperatures were taken for 0.25 kg of a substance cooling in surroundings which were at a steady temperature. During the cooling the substance solidified

t/min	0	1	2	3	4	5
θ/°C	82.0	80.0	78.0	76.1	74.2	72.7

t/min	6	30	31	32	33	34
θ/°C	72.4	72.4	71.3	67.8	63.6	59.7

Draw graphs of θ against t for the first 7 minutes and the last 7 minutes and from them estimate what the slopes of these two parts of cooling curve would have been at 72.4°C if the substance had not solidified. These are the rates of fall of temperature which the liquid and the solid would have had at 72.4°C.

If the s.h.c. of the substance in the liquid phase was 1850 J K^{-1}, what was (a) its s.h.c. in the solid phase (b) its rate of loss of energy at 72.4°C (c) its s.l.h. of fusion, assuming that it was solidifying for 26 minutes?

9-60 When a 100 W heater was embedded in some substance in the solid phase the temperature rose to 70°C and remained steady for 4.0 minutes before rising further. Later, with the heater switched off, the temperature fell to 70°C and remained steady for 16 minutes before falling further.

(a) What can you conclude from these readings?
(b) If the mass of substance was 80 g, what was its s.l.h. of fusion?

9-61 Discuss the effects on everyday life if we lived in a universe in which the values for water of the following quantities were half their present value: (a) the s.h.c. (b) the s.l.h. of fusion (c) the s.l.h. of vaporisation.

9-62 In the liquid phase benzene has a density of 800 kg m^{-3} at its n.b.p. (353 K) and in the vapour phase at that temperature it has a density of 2.69 kg m^{-3}. What is the volume of 0.100 kg of benzene at 353 K (a) in the liquid phase (b) in the vapour phase?
The work done by a substance when its volume increases by ΔV is $p\Delta V$, where p is the external pressure.
(c) What is the work done when 0.100 kg of benzene evaporates at 353 K, and atmospheric pressure (101 kPa)? The s.l.h. of vaporisation of benzene at 353 K is 3.94 × 10^5 J kg^{-1}.
(d) How much energy is needed to evaporate 0.100 kg of benzene at this temperature?
(e) How do you explain the difference between your answers to (c) and (d)?
(f) Express your answer to (c) as a percentage of your answer to (d).

Energy transfer

9-63 *Heating* is defined as the exchange of energy between bodies as a result of a temperature difference. Describe six everyday examples of heating: two of your examples should be of heating by conduction, two examples of heating by convection, and two examples of heating by radiation. Make your examples as different as possible.

9-64 Suppose a metal bar is lagged well enough for escape of energy from its surface to be negligible. Initially the whole bar is at room temperature. One end of the bar is then exposed to the air, and the other is kept at a steady temperature of about 50 K above room temperature. Describe what happens to the temperature at different points along the bar. You should explain why the temperature of the bar does not become uniform, and why the final temperature of the exposed end of the bar is above room temperature.

9-65 Sketch graphs to show how the temperature varies with distance along a metal bar when one end is kept at a high temperature and the other end is exposed to the air in the room if (a) the bar is uniform and well lagged (b) the bar is uniform and not lagged (c) the bar is well lagged but tapers towards the cooler end.

9-66 The ends of a well-lagged bar are kept at temperatures of 80°C and 10°C. What is the rate of flow of energy along the bar when it is (a) made of copper, is 0.20 m long, and has a cross-sectional area of 4.0×10^{-4} m^2 (b) made of copper, is 0.20 m long, and has a cross-sectional area of 2.0×10^{-4} m^2 (c) made of copper, is 0.40 m long, and has a cross-sectional area of 2.0×10^{-4} m² (d) made of iron, is 0.20 m long, and has a cross-sectional area of 4.0×10^{-4} m²? (λ for copper = 395 W m⁻¹ K⁻¹, λ for iron = 80 W m⁻¹ K⁻¹.)

9-67 What is the rate of transfer of energy through a brick wall of area 14 m² and thickness 120 mm, when its faces are at temperatures of 8.0°C and 12°C? (λ for brick = 1.0 W m⁻¹ K⁻¹.)

9-68 The thermal conductivity of copper is five times that of iron. A copper bar and an iron bar of the same dimensions are placed end-to-end. Assuming that the junction offers no resistance to the flow of energy, what will be the junction temperature when the free end of the iron bar is kept at 12°C and the free end of the copper bar is kept at 84°C?

If everything else is unchanged, but the length of the iron bar is reduced to one-third of its original value, what will then be the temperature of the junction?

9-69 A pond has a surface area of 30 m² and a layer of ice which is increasing in thickness at a rate of 1.35 mm per hour when it is 10 mm thick. If the upper surface of the ice is at − 0.50°C, and the water below the ice is at 0°C, calculate (a) the rate, in kg s⁻¹, at which ice is being formed (b) the rate of transfer of energy through the ice (c) the thermal conductivity of ice. (Density of ice = 920 kg m⁻³.)

9-70 A hot water storage cylinder in a house has a surface area of 1.6 m² and contains water at 70°C. Assume that the outer surface of the tank is also at 70°C. Room temperature is 20°C. If the rate of loss of energy from the tank can be analysed by assuming that there is a layer of still air 20 mm thick between the tank and the air of the room (with, therefore, a 50 K temperature difference across it) calculate the rate of loss of energy from the tank. (λ for air = 0.25 W m⁻¹ K⁻¹.)

An insulating jacket of glass wool ($\lambda = 0.040$ W m⁻¹ K⁻¹) of thickness 50 mm is then placed round the tank. The same layer of still air may be assumed to be next to the outside of the jacket. Calculate the temperature of the interface between the jacket and the layer of still air, and hence find the new rate of loss of energy from the tank. Assume that the effective cross-sectional area of the jacket can be taken to be 1.6 m².

9-71 Part (a) of the figure shows an electrical circuit in which a p.d. of 60 V is maintained across two resistors JK and KL. L is earthed, i.e. its potential is zero.
(a) What is the potential at K?
(b) If another resistor of resistance $2R$ is connected between K and L what is now the resistance between K and L?
(c) What is now the potential of K?

In part (b) two bars of the same material and same cross-sectional area A but of lengths l and $2l$ are placed end-to-end in good thermal contact. The ends of the composite bar are maintained at temperatures of 60°C and 0°C.
(d) What is the temperature at K?
(e) If a second bar of length $2l$ and area A is placed alongside the first in such a way that energy can flow without hindrance from JK into the double bar, what length of bar, of area A, is the double-bar equivalent to?
(f) What is now the temperature at K?

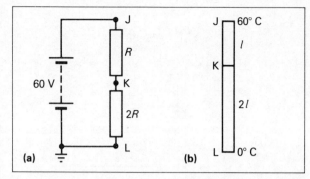

(a) **(b)**

9-72 Suppose a layer of ice has just formed on the surface of a pond. If the air temperature remains constant and below 0°C, would you expect the layer of ice to increase uniformly in thickness with time? Explain.

9-73 The figure shows a lagged iron pipe which carries hot water. The dashed circles are drawn at radii of 50 mm and 100 mm. The length of the pipe is 3.0 m and the external radius of the lagging is 150 mm. If the rate of escape of energy from the lagging is 18 W, what is the outward rate of transfer of energy across the circumference of (a) radius 50 mm (b) radius 100 mm?

If the thermal conductivity of the lagging is 0.035 W m⁻¹ K⁻¹, what is the temperature gradient $d\theta/dr$ in the lagging at a radius of (c) 50 mm (d) 100 mm? Give a *physical* reason to explain why your answer to (d) is half your answer to (c).

Sketch a graph to show the variation of $d\theta/dr$ with r, if the pipe has an external radius of 20 mm.

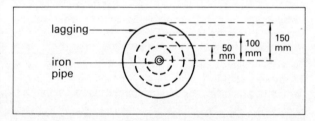

9-74 In a common method of measuring the thermal conductivity of a good conductor a specimen in the form of a metal bar is used. It is heated at one end and cooled at the other.
(a) Why must the bar be well lagged?
(b) In some forms of the apparatus (with electrical heating at the hotter end) the rate of flow of energy along the bar can be found without needing to make measurements on the cooling water. Why must the end of the bar still be cooled?
(c) Why should readings not be taken until all the temperatures are steady?
(d) Why would thermocouples be preferred to mercury thermometers for measuring the difference in temperature between two points along the bar?

9-75 In an experiment of the kind described in the previous question where a copper bar was used the following readings were taken:

temperatures at two points 100 mm apart on the bar: 73.2°C, 58.6°C

temperatures of the cooling water, at inlet and outlet: 12.7°C, 34.2°C

mass of cooling water collected in 10.0 minutes: 0.435 kg

diameter of bar: 38 mm

What value do these readings give for the thermal conductivity of copper?

9-76 In one form of an experiment to measure the thermal conductivity of a poorly conducting material a disc of the material is used. It is sandwiched between two brass plates of the same cross-sectional area; these brass plates carry thermometers. The upper brass plate has a steam jacket fixed to it, and this heats it. Energy then passes through the specimen to the lower brass plate, and thence into the surrounding air.

(a) Is it justifiable to assume that the thermometers measure the temperature of the surfaces of the specimen?

(b) Why is a specimen of this shape chosen, i.e. one which has a short length and large area?

(c) By what mechanism is energy transferred to the surrounding air?

(d) Is there any point in lagging any part of the apparatus?

(e) Why should the thermometer readings be recorded only when they are steady?

One way of measuring the rate of flow of energy through the specimen is to perform a second experiment. The specimen is removed, and the steam jacket and upper brass plate used to heat the lower brass plate to a temperature higher than the steady temperature θ_2 it reached during the first experiment. The jacket and upper brass plate are then removed, the specimen replaced, and readings of the lower brass plate's temperature taken as it cools. A cooling curve is plotted, and a tangent drawn at θ_2.

(f) What other measurements or values are needed so that the rate of loss of energy from the lower brass plate can be calculated?

(g) Is it justifiable to assume that this rate of loss of energy is the same as the rate of loss of energy from this plate during the main experiment?

9-77 In an experiment like that described in question **9-76** the following measurements were made:

temperatures of brass plates: 95.8°C, 92.1°C

thickness and diameter of polythene specimen: 0.34 mm, 106 mm

gradient of cooling curve at 92.1°C: 5.8 K min^{-1}

mass of lower brass plate: 0.890 kg

s.h.c. of brass: 370 J K^{-1} kg^{-1}.

What value do these measurements give for the thermal conductivity of polythene?

9-78 In another form of an experiment to measure the thermal conductivity of a poorly conducting material two identical discs of the material are used, with an electrical heating box sandwiched between them. Identical brass plates each carrying a thermometer are then placed on the outside of each specimen and the whole is clamped together and lagged round its circumference. The surfaces of the brass plates are left unlagged so that energy can escape from them. When rubber specimens of thickness 0.36 mm and diameter 100 mm were used the following measurements were made:

current flowing into heater: 7.6 A

p.d. across heater terminals: 6.2 V

average temperature of brass plates: 92.6°C

temperature of heating box: 102.9°C

What value do these measurements give for the thermal conductivity of rubber?

9-79 The thermal conductance U of a slab of material is defined by $U = \lambda/l$ where l is the thickness of the slab.

(a) What are the units of U?

(b) Explain what U measures.

The thermal resistance R of a slab of material is defined by $R = 1/U$. When two slabs of thermal resistance R_1 and R_2 are placed in series (e.g. a layer of plasterboard fixed to a brick wall) the total thermal resistance R is given by $R = R_1 + R_2$, as with electrical resistance.

(c) Calculate U, and hence R, for each of the following: (i) brick 120 mm thick ($\lambda = 1.0$ W m^{-1} K^{-1}) (ii) breeze block 120 mm thick ($\lambda = 0.40$ W m^{-1} K^{-1}) (iii) foamed plastic 80 mm thick ($\lambda = 0.050$ W m^{-1} K^{-1}) (iv) glass 4.0 mm thick ($\lambda = 1.0$ W m^{-1} K^{-1}) (v) air 25 mm thick ($\lambda = 0.25$ W m^{-1} K^{-1})

(d) What is the total thermal resistance of the following combinations of the above, when placed in series: (i) air + brick + air (ii) air + brick + foamed plastic + brick + air (iii) air + glass + air?

(e) What are the values of U for each of these three combinations?

(f) What is the rate of transfer of energy through a wall which measures 5.0 m by 3.0 m, when the outside and inside temperatures are 20°C and 2°C, and it has a thermal conductance of (i) 3.1 W m^{-2} K^{-1}, (ii) 0.45 W m^{-2} K^{-1}?

(g) What is the rate of transfer of energy through a window which measures 2.0 m by 1.0 m, when the outside and inside temperatures are 20°C and 2°C, and it has a thermal conductance of 4.9 W m^{-2} K^{-1}?

9-80 In part (g) of question **9-79** you calculated the rate of transfer of energy through a window—treated as a layer of glass sandwiched between two layers of still air, the outer surfaces of which were at 20°C and 2°C. Using the data given in that question, calculate the difference in temperature between the two surfaces of the glass, and hence find the temperatures of the surfaces of the glass.

9-81 The Building Regulations in Great Britain state that for new houses the U-value (thermal conductance) for walls must be less than 1.0 W m^{-2} K^{-1}. A typical cavity wall consists of two brick walls, each 120 mm thick, with a layer of air 50 mm thick between them. Does this type of construction meet the regulations? [Hint: calculate U for each layer, including that for a layer of still air about 25 mm thick on the outsides of the brick walls, and hence find R for each layer, and hence the total thermal resistance.] (Values of λ: brick 1.0 W m^{-1} K^{-1}, air 0.25 W m^{-1} K^{-1}.)

9-82 If the U-value (see question **9-79**) for a ceiling with an open loft above it is 0.43 W m^{-2} K^{-1}, and it is then covered wth a layer of aluminium foil which has a U-value of 0.21 W m^{-2} K^{-1}, what is the final U-value? Why would the foil be placed with its shiny side uppermost?

9-83 Compare the ways in which a room is heated by (a) a radiator which is part of a central heating system (b) a coal fire in an open grate.

9-84 When an electric lamp is switched on it very quickly becomes white-hot but although energy continues to be supplied from the mains the lamp does not become any hotter. Why is this? What decides the temperature which the lamp reaches?

9-85 An office is heated by an electric heater which is switched on at 9 a.m. The temperature of the office begins to rise but by 11 a.m. it is found that the temperature has stopped rising, although the heater is still on. Why is this? On a colder day it is found that the maximum temperature is lower, and is reached by 10.30 a.m. Explain this.

9-86 An immersion heater fitted to a hot-water tank will always be fitted at the bottom, so that the circulation of water by natural convection will ensure that all the water is uniformly heated. Some tanks, however, have a second heater fitted about one-quarter of the way down from the top. Suggest a reason for this.

9-87 Refrigerators always have their freezing compartments at the top. Why is this?

9-88 The figure shows a *heat sink* which can be attached to a piece of high-power electrical apparatus so that it does not get too hot. Explain its shape, and why it is usually painted black. This heat sink is said to have 'a thermal resistance of

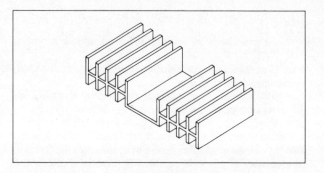

4 K W^{-1} with fins vertical in free air.' Consider what the units used in this statement imply, and explain what the statement means. What would happen if the heat sink were bolted to a device in which energy was being generated at a rate of 20 W?

9-89 Two copper spheres A and B hang in identical enclosures. They are initially at the same temperature, which is higher than that of the enclosure. If the diameter of A is twice that of B, what are, initially, the ratios (a) (rate of loss of energy of A)/(rate of loss of energy of B) (b) (rate of fall of temperature of A)/(rate of fall of temperature of B)? In a cold winter why are small birds more likely to die than large birds?

9-90 A hot metal sphere is suspended in air by a nylon fibre: assume that the loss of energy by conduction along the fibre is negligible. Describe (quantitatively, if possible) and explain the factors on which the initial rate of loss of energy to the surroundings depends.

10 Thermometers and the ideal gas

Data g = 9.81 N kg^{-1}
Avogadro constant $L \times 6.02 \times 10^{23}$ mol^{-1}
molar gas constant R = 8.31 J K^{-1} mol^{-1}
Boltzmann constant k = 1.38 × 10^{-23} J K^{-1}
triple point of water = 273.16 K
atmospheric pressure = 101 kPa
Relative molar masses M_r: hydrogen 2, helium 4, nitrogen 28, oxygen 32.

Measuring temperature

10-1 Two copper blocks, placed together in good thermal contact, will soon reach the same temperature. But if one of the blocks has a greater mass than the other, the blocks will not have the same amount of internal energy. What is it that is the same for both?

10-2 (a) The length of a mercury column in a mercury thermometer is 15.3 mm at the ice point and 47.8 mm at the steam point. What is the temperature on the centigrade scale of this thermometer when the length of the thread is 21.6 mm?
(b) The resistance of a piece of platinum is 3.254 Ω at the ice point and 4.517 Ω at the steam point. It is used to measure the same temperature as in (a) and its resistance is then 3.494 Ω. What is then the temperature on the centigrade scale of *this* thermometer?
(c) Comment on the discrepancy between your answers to (a) and (b).

10-3 Does a thermometer read its own temperature or the temperature of its surroundings?
 Suppose a mercury thermometer was put in a place shaded from direct sunlight, and reached a steady temperature. Another identical thermometer placed in direct sunlight would give a higher steady temperature. What are these two thermometers measuring? Explain, making reference to the rates of emission and absorption of energy by the bulbs, why the readings are different.

10-4 In a mercury thermometer why is (a) the bore narrow (b) the bulb relatively large (c) the glass of the bulb thin (d) mercury preferred to other liquids (e) the inside evacuated?

10-5 Three sources of error in a simple form of constant-volume gas thermometer are (a) there is a *dead space* between the heated bulb and the fixed mercury level (b) the material of the bulb expands as its temperature rises (c) the density of the mercury will change if the manometer is near the bulb which is being heated. For each source of error discuss whether the measured temperature will be too high or too low.

10-6 At low pressures the ratios of the pressure of a gas, kept at constant volume, at the triple point of water and the melting point of tin, is 1.849. What is the temperature of the m.p. of tin?

10-7 In a measurement of the n.b.p. of mercury using a gas thermometer three separate measurements were made, each starting with a different amount of gas in the thermometer, and consequently a different initial pressure. At the triple point of water these three pressures were 75.000 kPa, 50.000 kPa and 25.000 kPa. When the bulb was at the temperature of the boiling mercury these three pressures became 172.969 kPa, 115.303 kPa and 57.646 kPa respectively. Calculate, to six significant figures, the three values for the n.b.p. of mercury which these three pairs of readings give, and hence calculate, to five significant figures, the n.b.p. of mercury as measured by this gas thermometer.

10-8

	triple point (273.16 K)	room temperature
Resistance of resistance thermometer/Ω	58.765	62.388
Pressure recorded by constant-volume gas thermometer/kPa	10.821	11.479

Use the data to calculate room temperature on the scale of each thermometer (a) on the kelvin scale (b) on the Celsius scale.

10-9 What type of thermometer would you use to measure each of the following? In each case explain your choice:
(a) the temperature of the filament of an electric light bulb
(b) the temperature just after ignition in a cylinder of an internal-combustion engine
(c) the n.m.p. of titanium (1950 K)
(d) the b.p. of water on a mountain
(e) the temperature difference between the inlet and outlet temperatures in a continuous flow method for measuring the s.h.c. of a liquid
(f) the n.b.p. of argon (84 K).

10-10 Comment on this statement: 'It is better to use mercury rather than alcohol in a liquid-in-glass thermometer because mercury expands uniformly with temperature.'

10-11 One junction of an iron-constantan thermocouple is kept at the ice point. When the other junction is at the steam point the e.m.f. is 5.268 mV. When the second junction is at a point where an ideal gas thermometer reads 50.00°C the e.m.f. is 2.585 mV. What is the temperature on the Celsius scale of the thermocouple thermometer? Comment on your answer.

Boyle's law and the ideal gas

10-12 In a simple experiment to verify Boyle's law the length l of a fixed mass of air in a uniform tube was measured, its temperature being kept constant. The pressure of the air was measured with a mercury manometer, the height of the mercury in the arm connected to the air being h_1, and the height of the arm open to the atmosphere being h_2. Atmospheric pressure was able to support 755 mm of mercury. The table gives the readings taken.

l/mm	45	53	63	75	84
h_1/mm	385	462	534	593	627
h_2/mm	681	602	530	471	435

Calculate the pressure of the air in each of the five cases, and draw a graph which will enable you to verify whether Boyle's law is true for this gas. Also check whether the law is true *without* drawing a graph. Which do you think is the better method?

10-13 At the beginning of a journey the pressure of the air in a car tyre is 276 kPa and the temperature is 12°C. After being driven the pressure is 303 kPa. Assuming that the volume of the air remains constant, what is now the temperature?

10-14 On a day when the atmospheric pressure supports a column of mercury 758 mm high, the height of the mercury in a faulty barometer (which contains air in the space above the mercury) is 725 mm, and the height of the space above the mercury is 153 mm. When the barometer tube is raised vertically the height of the mercury becomes 730 mm. Calculate (a) the new height of the space above the mercury (b) the distance the barometer tube was raised.

10-15 At a certain temperature 1.6×10^{-3} mol of an ideal gas occupies a volume of 3.0×10^{-5} m³ and exerts a pressure of 120 kPa. If the temperature remained unchanged what would be the pressure if, separately, (a) the volume became 4.5×10^{-3} m³ (b) the volume became 2.5×10^{-3} m³ (c) the amount of gas was 2.0×10^{-3} mol (d) the number of molecules of the gas was halved?

10-16 Some gas occupies a volume of 6.0×10^{-3} m³ and exerts a pressure of 80 kPa at a temperature of 20°C. What pressure does it exert if, separately, (a) the temperature is raised to 40°C (b) the volume is halved (c) the temperature is raised to 586 K (d) the volume becomes 2.6×10^{-2} m³ (e) the volume becomes 7.7×10^{-3} m³ and the temperature becomes 57°C?

10-17 An air bubble of volume 3.0×10^{-5} m³ escapes from a driver's equipment at a depth of 45 m where the water has a temperature of 5.0 °C. What is its volume as it reaches the surface, where the temperature is 12°C? (Atmospheric pressure = 101 kPa, density of seawater = 1020 kg m⁻³.)

10-18 A vessel of volume 5.0×10^{-2} m³ contains an ideal gas at a pressure of 120 kPa and a temperature of 350 K.

How many moles of gas are present? What mass of gas is present if it is (a) hydrogen (b) oxygen?

10-19 If the following masses of gas, all at the same temperature, are placed successively in the same container, which will exert the greatest pressure, and which the least: (a) 20 g of hydrogen (b) 20 g of helium (c) 200 g of oxygen?

10-20 The best man-made vacuum has a pressure of the order of 10^{-14} Pa. In a volume of 1.0 m^3 at a temperature of 300 K how many (a) moles (b) molecules are there?

10-21 Hydrogen at 300 K and a pressure of 101 kPa contains 2.4×10^{25} molecules per cubic metre. How many molecules per cubic metre will there be at a place where the temperature is 400 K and the pressure is 1.01 kPa?

How many molecules of helium would there be per cubic metre at a temperature of 400 K and a pressure of 1.01 kPa?

10-22 A vessel of volume 0.20 m^3 contains a mixture of (i) 2.0 g of hydrogen molecules (ii) 8.0 g of helium molecules, at a temperature of 320 K.
(a) Calculate the number of moles of (i) hydrogen (ii) helium.
(b) What is the total amount of substance?
(c) What is the pressure in the vessel?

10-23 Two flasks of volume 2.00 litres and 3.00 litres respectively are connected by a tube of negligible volume and contain air at a pressure of 120 kPa. Initially both flasks are at 20°C. The 3.00 litre flask is then heated to 100°C.
(a) Calculate the original total number of moles of air in the two flasks.
(b) Writing p for the final pressure in the system, write down an expression involving p for (i) the number of moles in the 2.00 litre flask (ii) the number of moles in the 3.00 litre flask, after the heating.
(c) Use the idea that the total number of moles remains constant to calculate the final pressure p.

10-24 A car tyre contains 1.60×10^{-2} m^3 of air at a pressure of 300 kPa and a temperature of 18°C. A pump of volume 2.00×10^{-4} m^3 contains air at atmospheric pressure and a temperature of 18°C. The pump is used 20 times to force air into the tyre. Assuming that there is no rise in temperature of any part of the system, (a) what was the original number of moles of air in the tyre? (b) what is the number of moles supplied by the pump? (c) what is the final pressure in the tyre?

10-25 Consider a rectangular box with sides of 0.30 m, 0.40 m, 0.50 m. Suppose it contains 1.5×10^{24} molecules, each of mass 5.0×10^{-26} kg. Suppose that each face of the box has one-third of the molecules moving at right-angles to it, and that all the molecules have the same speed of 500 m s^{-1}. Consider the face which measures 0.30 m $\times$ 0.40 m. Calculate (a) the time between successive impacts of a particular molecule on this face (b) the size of the change of momentum when a molecule strikes this face (c) the average rate of change of momentum at this face, caused by one molecule (d) the average force, caused by all the molecules, on this face (e) the pressure on this face.

Why would a similar calculation, for another face, give the same value for the pressure?

10-26 The density of argon is 1.61 kg m^{-3} at a pressure of 100 kPa. What is the r.m.s. speed of the argon atoms under these conditions?

What would be the r.m.s. speed if the pressure were halved, the temperature remaining the same?

10-27 How does the simple kinetic theory of gases explain why the pressure of a fixed mass of gas at constant temperature is inversely proportional to its volume?

10-28 How does the simple kinetic theory of gases explain the following:
(a) the volume of a gas is proportional to its kelvin temperature, at constant pressure
(b) the pressure of a gas is proportional to its kelvin temperature, at constant volume?

10-29 Use the equations $p = \frac{1}{3}\rho\overline{c^2}$ and $pV = \frac{m}{M}RT$ to derive an expression for $\overline{c^2}$ in terms of the molar mass M [$= (M_r/1000)$ kg mol^{-1}], the molar gas constant R, and the kelvin temperature T. Calculate the r.m.s. speed at 300 K of the molecules of (a) hydrogen (b) helium (c) carbon dioxide ($M_r = 44$).

Why would you expect the speed of sound in a gas at a certain temperature to be comparable with, but less than, r.m.s. speed of the molecules of that gas at that temperature?

10-30 The measured speeds of 10 vehicles on a motorway are, in m s^{-1}, 31, 32, 28, 40, 33, 32, 35, 34, 32, 25. Calculate (a) their mean speed (b) their r.m.s. speed.

10-31 A vessel of volume 0.075 m^3 contains helium at a temperature of 290 K. If the pressure of the gas is 102 kPa, calculate (a) the mass of helium (b) the r.m.s. speed of the helium atoms.

10-32 Air at s.t.p. has a pressure of 101 kPa and a temperature of 273 K. Take the mean value of M_r for air to be 30. Deduce (a) the r.m.s. speed of an air molecule (b) the mass of air in a volume of 1.0 litre (c) the number of molecules in a volume of 1.0 litre.

Consequences of kinetic theory

10-33 Some molecules have a mean translational k.e. of (a) 5.65×10^{-21} J (b) 1.24×10^{-20} J. What is the temperature of the gas of which they form part?

10-34 What is the mean translational k.e. at 290 K of the molecules of the following gases: (a) hydrogen (b) nitrogen (c) bromine ($M_r = 160$)?

10-35 What is the r.m.s. speed at 290 K of the molecules of each of the gases listed in question **10-34**?

10-36 A vessel contains a mixture of hydrogen and oxygen gases. Will the molecules of the two gases have (a) different mean translational kinetic energies (b) different r.m.s. speeds? Explain

10-37 The oil droplets used in Millikan's experiment have a typical mass of 5.0×0^{-15} kg. What is their r.m.s. speed in air at a temperature of 290 K? When you do the experiment is there any evidence of this motion?

10-38 A vessel contains 2.0 g of hydrogen, 2.0 g of helium and 10 g of neon. What contribution does each gas make to the pressure of 100 kPa in the vessel? (For neon $A_r = 20$.)

10-39 0.93% by volume of the Earth's atmosphere is argon. What is (a) the number of moles of argon in 100 mol of atmosphere (b) the partial pressure of the argon when atmospheric pressure is 100 kPa?

10-40 On a day when the temperature is 287 K and the relative humidity 60% the partial pressure of the water molecules of water vapour in the air is 0.96 kPa. If atmospheric pressure is 101.00 kPa, what is (a) the pressure caused by the air molecules (b) the number of water vapour molecules as a percentage of the total number of molecules present?

When using a barometer to find atmospheric pressure do you need to make a correction for the presence of water vapour molecules in the air?

10-41 A previously evacuated vessel of volume 1.0×10^{-2} m^3 contains 1.0 mol of liquid N_2O_4. When its temperature is raised all the liquid evaporates, and at 32°C 40% of the N_2O_4 molecules have dissociated into NO_2 molecules, in accordance with the equation $N_2O_4 \rightleftharpoons 2NO_2$. Find (a) the number of moles of gas now present in the vessel, and (b) the pressure in the vessel.

10-42 Neon gas contains atoms of different mass: 90.9% have a mass number of 20, and 8.8% have a mass number of 22. If some of the gas is placed in a container which has a hole fitted with a plug of porous material through which the gas can diffuse into an evacuated space, find the ratio r_{20}/r_{22} of the initial rates of diffusion of these two isotopes.

If this process were to be used as a means of separating the isotopes of neon, why would it be important not to let the process continue for very long?

Uranium contains 99.3% of atoms of mass number 238, and 0.7% of atoms of mass number 235. Combined with fluorine ($A_r = 19$) it forms the gaseous compound UF_6. Why can a diffusion process used with UF_6 to separate the isotopes of uranium not be expected to be as effective as that described for neon?

10-43 An early method of investigating the distribution of molecular speeds used apparatus like that shown in the figure. Atoms or molecules emerged from the oven and passed through the slit S and into a rotating drum. They crossed the drum and were deposited on a glass plate. The intensity of the deposition was measured optically.
(a) Draw a diagram of the glass plate, labelled AB as in the figure, and shade it to show what the distribution of atoms or molecules would be like.
(b) At which end are the fastest molecules? Explain.
(c) What is the r.m.s. speed of a bismuth atom at 1100 K? ($A_r = 209$, $m_u = 1.66 \times 10^{-27}$ kg)
(d) If the diameter of the drum is 200 mm, how fast must it be rotated if atoms with the r.m.s. speed are to strike the glass plate 45° round the circumference from A?

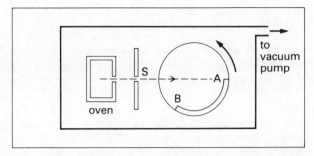

10-44 The figure shows the distribution of moelcular speeds v for 1 000 000 oxygen molecules at two temperatures. At any particular whole-number speed the height of the curve represents the number of molecules which have a speed within ± 0.5 m s^{-1} of that speed (e.g. at 300 K about 2000 molecules have a speed between 319.5 m s^{-1} and 320.5 m s^{-1}).
(a) Estimate the number of oxygen molecules which have speeds within ± 0.5 m s^{-1} of 750 m s^{-1} at (i) 300 K (ii) 600 K.
(b) Is the area between the 300 K curve and the v-axis the same as the area between the 600 K curve and the v-axis? What does this area represent?
(c) Estimate the number of oxygen molecules which have speeds of less than 250 m s^{-1} at (i) 300 K (ii) 600 K.

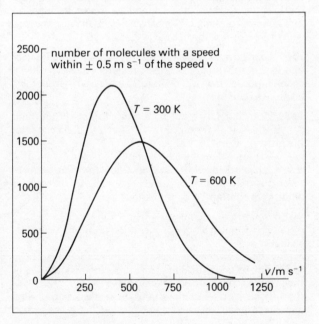

The internal energy of a gas

10-45 How much internal energy is there in 2.00 mol of an ideal gas ($C_V = \frac{3}{2} R$) at a temperature of (a) 300 K (b) 600 K?

If 2.00 mol of an ideal gas at 600 K are given 14.9 kJ of energy by heating at constant volume, what does the temperature become?

10-46 An ideal gas has a molar heat capacity (at constant volume) of $\frac{3}{2}R$.

(a) How much energy is needed to raise the temperature of 10.0 mol of an ideal gas from 273 K to 500 K at constant volume?

(b) If 2.50 kJ of energy are given to 5.00 mol of an ideal gas at constant volume, what will be its rise in temperature?

10-47 Hydrogen and nitrogen have s.h.c.s of 1.018×10^4 J K^{-1} kg^{-1} and 745.0 J K^{-1} kg^{-1} respectively, and their molar masses M_{m} are 2.016×10^{-3} kg mol^{-1} and 28.01×10^{-3} kg mol^{-1} respectively. What are their molar heat capacities? Comment on your result.

10-48 The m.h.c. of hydrogen (at constant volume) is roughly 12.5 J K^{-1} mol^{-1} at low temperatures, roughly 20.8 J K^{-1} mol^{-1} between 300 K and 800 K, and roughly 29.1 J K^{-1} mol^{-1} at high temperatures.

(a) What does this information tell us about the amount of energy needed to raise the temperature of some hydrogen (e.g. 2.0 mol through 10 K) in different temperature ranges?

(b) What molecular explanation can you give for the values of the m.h.c. being different?

10-49 Refer to the previous question. If the cylinder is thermally isolated, so that no energy can enter or leave by heating, what is the change of internal energy in each case?

10-50 Refer to question **10-48**. If the cylinder makes good thermal contact with its surroundings, so that its temperature remains constant, how much energy do the surroundings gain or lose by heating in each case?

10-51 Explain in molecular terms why (a) the internal energy of a gas in a thermally isolated container increases when a piston is pushed inwards, compressing the gas (b) work must be done to compress a gas.

10-52 A cylinder is fitted with a frictionless piston, and contains gas at a constant pressure of 50 kPa. The area of the piston is 1.0×10^{-2} m^2.

(a) What force does the gas exert on the piston?

(b) What work does the gas do when the piston moves reversibly outwards 5.0 mm, the pressure inside the cylinder being kept constant, e.g. by heating?

10-53 Some gas is contained in a cylinder fitted with a piston. How much work is done in the following reversible changes:

(a) by the gas, when it pushes the piston back and increases its volume by 6.0×10^{-3} m^3? The pressure on both sides of the piston is constant and equal to 120 kPa

(b) on the gas, when the atmosphere pushes the piston inwards to reduce the volume by 2.5×10^{-3} m^3. The pressure on both sides of the piston is constant and equal to 100 kPa?

10-54 2.5 mol of an ideal gas ($C_V = \frac{3}{2}R$) has its temperature raised from 250 K to 450 K (a) with its volume kept constant (b) with its pressure kept constant, so that it must expand (c) with neither its pressure nor its volume kept constant. In each case how much internal energy does it gain?

In which of these cases can you calculate how much energy has been supplied by heating?

How much was supplied?

10-55 With the usual sign convention 0.50 mol of an ideal gas ($C_V = \frac{3}{2}R$) undergoes a reversible process at a constant pressure of 80 kPa for which $\Delta W = -500$ J and $\Delta Q = +250$ J.

(a) What do the *signs* attached to the values of ΔW and ΔQ tell us?

(b) What is the change in internal energy of the gas?

(c) What is the change in temperature of the gas?

(d) What is the change of volume of the gas?

10-56 0.250 mol of an ideal gas ($C_V = \frac{3}{2}R$) is enclosed in a cylinder fitted with a frictionless piston and is heated reversibly so that its temperature rises from 288 K to 320 K. The pressure on both sides of the piston remains constant at 102 kPa.

(a) What is the change of internal energy of the gas?

(b) What were the original and final volumes of the gas, and what is the change in volume?

(c) How much work is done by the gas?

(d) How much energy is supplied by heating?

10-57 Refer to the previous question. This was a constant pressure process in which you now know ΔQ, n, ΔT. Calculate the constant-pressure m.h.c. C_p.

10-58 Explain why the constant pressure molar heat capacity is greater than the constant volume molar heat capacity. Are there more than two values of molar heat capacity?

10-59 0.450 mol of an ideal gas ($C_V = {}^3/_2 R$) is contained in a cylinder fitted with a frictionless piston. Its initial temperature is 283 K, and its pressure is 90.0 kPa.

(a) What is its initial volume?

(b) It is heated reversibly so that its temperature rises to 326 K, its pressure remaining constant. What is now its volume?

(c) What are ΔU, ΔW, ΔQ for the gas?

(d) What is the m.h.c. for the gas under these conditions?

(e) Comment on the fact that your answer to (d) is neither C_V ($= 12.5$ J K^{-1} mol^{-1}) nor C_p ($= 20.8$ J K^{-1} mol^{-1}).

10-60 0.30 mol of an ideal gas ($C_v = \frac{3}{2}R$, $C_p = \frac{5}{2}R$) is enclosed in a cylinder fitted with a frictionless piston and heated reversibly so that its temperature rises from 275 K to 345 K, its pressure remaining constant.

(a) How much energy is supplied by heating?

(b) What is the change in internal energy?

(c) Is work done on, or by, the gas, and how much work is done?

10-61 An ideal gas expands reversibly and isothermally (i.e. its temperature remains constant). It therefore obeys Boyle's law. Initially the pressure is 240 kPa and the volume 1.0×10^{-3} m^3. Calculate its volume for pressures p given by $p/$kPa $= 210, 180, 150$ and 120 and draw a graph of p (on the y-axis) against V (on the x-axis). Hence calculate the work done by the gas when it expands from a pressure of 240 kPa to a pressure of 120 kPa.

Isothermal and adiabatic processes

10-62 How much work is done by an ideal gas when it changes reversibly from a pressure of 100 kPa and a volume of 3.0×10^{-3} m³ to (a) a volume of 5.0×10^{-3} m³ at constant pressure (b) a pressure of 120 kPa at constant volume?

10-63 An ideal gas expands isothermally and reversibly (i.e. obeying the law pV = constant) from a volume of 1.0×10^{-3} m³ and a pressure of 240 kPa to a volume of 3.0×10^{-3} m³. Plot a graph of p against V for this expansion, using values of V at intervals of 0.5×10^{-3} m³. Deduce the work done by the gas as it expands (it is represented by the area between the graph and the volume-axis). What is (a) the change of internal energy of the gas (b) the energy supplied to the gas by heating?

10-64 0.40 mol of an ideal monatomic ($C_v = 12.5$ J K⁻¹ mol⁻¹) at a pressure of 100 kPa and a temperature of 300 K is taken round the following cycle: (i) it is cooled at constant pressure until it is at 150 K (ii) it is heated at constant volume until its temperature is 450 K (iii) it is heated at constant pressure until it regains its original volume (iv) it is cooled at constant volume until it is in its original state.

Calculate the original volume of the gas, and sketch the cycle on a p–V diagram, marking in the values of pressure, volume and temperature. Calculate (a) the highest temperature reached (b) ΔU, ΔW and hence ΔQ for each of the processes (i) to (iv) (c) the net work done in the cycle (d) the energy supplied by heating during processes (ii) and (iii) (e) the fraction of the energy supplied to the gas which is converted to mechanical energy.

Comment on the fact that ΔU for the whole cycle is zero.

10-65 An ideal gas is taken reversibly along the path abc in the figure. The work done by the gas is 40 kJ, and the energy supplied by heating is 90 kJ. When it is taken along path adc the work done is 20 kJ.
(a) Describe the changes which occur when the gas is taken along (i) abc (ii) adc.
(b) What is the pressure ratio p_2/p_1?
(c) If U_a = 40 kJ, what is U_c?
(d) How much energy is supplied by heating along path adc?

(e) If U_d = 70 kJ, what is ΔQ for the paths (i) ad, (ii) dc?
(f) If the work done on the gas when it returns along the curved path ca is 35 kJ, what is ΔQ for the path?
(g) What does the area of the rectangle $abcd$ represent?
(h) How much work would be done on the gas if it returned along a straight line joining c and a?

10-66 How can a gas be heated isothermally, i.e. without changing its temperature? Describe a practical arrangement in which this would be possible.

10-67 An ideal gas is (i) allowed to expand isothermally so that its pressure p_1 falls to a pressure p_2 and its volume V_1 increases to volume V_2, then (ii) compressed at constant pressure p_2 to its original volume V_1, then (iii) heated at constant volume V_1 until its pressure is again p_1. All the processes are reversible. Sketch a p–V graph for the cycle.
(a) Draw up a table with headings ΔW, ΔU, ΔQ and ΔT and for each of the processes (i), (ii) and (iii) complete the table to show whether these quantities are positive, negative or zero.
(b) For the complete cycle are the values of ΔW, ΔU, ΔQ and ΔT positive, negative or zero?

10-68 Describe the type of process, for which, with the notation of the first law of thermodynamics, (a) $\Delta Q = 0$ (b) $\Delta U = 0$ (c) $\Delta W = 0$.

10-69 An ideal gas is (i) compressed adiabatically from a pressure p_1 and volume V_1 to a pressure p_2 and volume V_2, then (ii) allowed to expand isothermally to a pressure p_3 and its original volume V_1. Its pressure then (iii) falls at constant volume to its original pressure p_1. All the processes are reversible. Sketch a p–V graph of the cycle.

(a) Draw up a table with headings ΔW, ΔU, ΔQ, and ΔT and for each of the processes (i), (ii) and (iii) complete the table to show whether these quantities are positive, negative, or zero.
(b) For the complete cycle are the values of ΔW, ΔU, ΔQ and ΔT positive, negative or zero?

10-70 2.5 mol of an ideal gas (for which $\gamma = 1.4$) expands adiabatically and reversibly from a volume of 0.050 m³ and a pressure of 150 kPa to a volume of 0.10 m³.

(a) What was the initial temperature?
(b) What was the final pressure?
(c) What was the final temperature?
(d) How much energy was supplied by heating?
(e) What was the change in internal energy?
(f) How much work was done by the gas?
For this gas the molar heat capacity $C_V = 20.8$ J K⁻¹ mol⁻¹.

10-71 A motor car tyre contains air ($\gamma = 1.4$) at a pressure of 300 kPa and a temperature of 15°C. If the air is allowed to expand reversibly and adiabatically until it reaches a pressure of 100 kPa, what will be the new temperature of the air?

Suppose the expansion had occurred through the bursting of the tyre. Would this have been a reversible process? Would the final temperature of the air have been higher or lower than your previously calculated value?

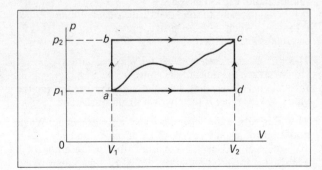

10-72 Explain what is meant by reversibility. Discuss to what extent the following processes are reversible: (a) the passage of an electric current through a resistor (b) the evaporation of water in a closed container placed on a hotplate (c) the 'recharging' of a car battery (d) loading and unloading a mass on a spring balance (e) the oscillations of the air in a sound wave.

11 Real gases and thermodynamics

Data $g = 9.81$ N kg^{-1}
   molar gas constant $R = 8.31$ J K^{-1} mol^{-1}
   atmospheric pressure = 101 kPa
   s.h.c. of water = 4.18 kJ K^{-1} kg^{-1}
   s.l.h. of fusion of ice = 0.334 MJ kg^{-1}
   s.l.h. of vaporisation of water = 2.26 MJ kg^{-1}

Real gases

11-1 Examine the figure, which shows a $p–V–T$ surface for a real substance, and where T_c and T_{tr} are the *critical temperature* and the *triple-point temperature*.
(a) Explain what these terms mean.
(b) Can the substance have more than one pair of values of p and V at (i) T_c (ii) T_{tr}?
(c) Why is it the triple point of water which is chosen as one of the fixed points of the ideal-gas temperature scale, and not the temperature at the critical point?

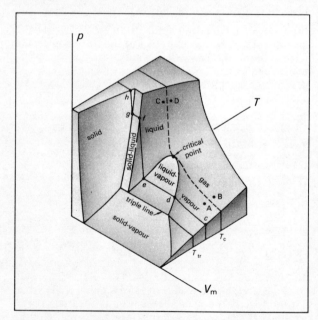

11-2 Look again at the previous figure. Suppose the substance were contained in a glass vessel, so that its condition could be observed. What, if anything, would you notice as the substance moved on the surface (a) from A to B (b) from C to D (c) from B to D (d) along the line *cdefgh* (e) starting from a point between *d* and *e* and was heated at constant volume so that it passed through the critical point?

11-3 At 20°C carbon dioxide is a gas or a vapour. Draw a $p–V$ graph to show what happens when the pressure on it is increased to very high values while the temperature remains constant. Also describe what is happening at different stages in the process.
   Is the $p–V$ graph for oxygen, at 20°C, similar? Is there any temperature at which the graph for oxygen is similar?

11-4 Explain, in molecular terms (a) what happens when a gas liquefies (b) why liquefaction happens at low temperatures (c) why the pressure of a vapour decreases while it is liquefying in a space of constant volume.

11-5 In an ideal gas the molecules merely repel each other when they collide; in a real gas there are forces of repulsion and attraction which vary continuously with the centre-to-centre separation of the molecules. Explain why, when a real gas expands into an evacuated space it cools, whereas an ideal gas does not. [Hint: what happens to (a) its total internal energy (b) its intermolecular potential energy (c) its translational kinetic energy?]

11-6 0.050 mol of a substance is contained in a vessel of volume 1.0×10^{-3} m^3 at a temperature of 290 K. The pressure in the vessel is 100 kPa. Verify that the substance is partly gaseous and partly liquid, and assuming that the gaseous part behaves as an ideal gas, calculate the amount of liquid present.

11-7 The triple point of carbon dioxide occurs at a pressure of 518 kPa and a temperature of 217 K: its critical pressure is 7360 kPa and its critical temperature is 304 K.
(a) Sketch a $p–T$ graph (which need not be to scale) for carbon dioxide, to show the boundaries between the solid, liquid and gas phases.
(b) What must be the minimum pressure in a cylinder containing liquid carbon dioxide?
(c) What happens to solid carbon dioxide if its temperature is raised when the external pressure is (i) 100 kPa (ii) 1000 kPa?

11-8 (a) Write down the Van der Waals equation.
(b) What are the units of the constants a and b?
(c) When $T = 0$ what does the equation predict for the value of V_m? What is the physical significance of b?

11-9 For butane the constants a and b in the Van der Waals equation are given by $a = 1.47$ Pa m^6 mol^{-1} and $b = 1.23 \times 10^{-4}$ m^3 mol^{-1}. Use the equation to plot the graph of p against V_m for a temperature of 390 K, and values of V_m

given by $V_m/10^{-4}$ m^3 mol^{-1} = 2.00, 2.25, 2.50, 3.00, 3.50, 4.00, 4.50, 5.00, 6.00, 7.00, 8.00, 9.00, 10.00.

Label the part of the graph where butane is liquid and the part where it is vapour, and estimate from the graph the range of values of V_m in which butane coexists in the liquid and vapour phases.

Vapour saturating a space

11-10 Refer to the figure for question **11-1**, which shows a p–V–T surface for a real substance. If you wanted to see how the s.v.p. varies with temperature for this substance, in which direction would you look? Sketch a graph to show this and the other boundary lines between the different phases which can be seen from this direction, and label (a) the axes of your graph (b) the triple point (c) the critical point (d) the regions between the boundary lines in which the substance is solid, liquid and vapour.

How would you describe the region where the temperature is higher than the critical temperature?

11-11 A closed vessel contains some liquid and the vapour of the liquid only. Describe, in molecular terms, (a) what is happening in the vessel (b) what will happen if the temperature is raised slightly.

11-12 Consider a vessel which contains only a few drops of liquid together with its own vapour. The volume of liquid is always negligible compared with the volume of the vessel. The vessel is fitted with a piston. Describe, in molecular terms, what happens when the piston is moved inwards so that the volume of the vessel is reduced by one-tenth. The temperature remains constant.

What is now (a) the pressure, if the original pressure was p_0 (b) the number of molecules in the vapour phase, if the original number was N_0?

11-13 A vessel contains air and water vapour, and a trace of water, which is enough to ensure that the space is saturated. The pressure in the space is caused by both the air and the water vapour: the pressure is the sum of the partial pressures of these two gases. What will happen to the partial pressure of (a) the air if (i) the volume is doubled (ii) the (kelvin) temperature is doubled (b) the water vapour if (i) the volume is doubled (ii) the (kelvin) temperature is doubled?

11-14 The table gives corresponding values of temperature and the s.v.p. of water.

θ/°C	0	5	10	15	20	25	30
s.v.p./kPa	0.61	0.87	1.23	1.71	2.34	3.17	4.25

Draw a graph of s.v.p. against temperature.

The *relative humidity* (r.h.) of an atmosphere is defined by the equation

$$\text{r.h.} = \frac{\text{partial pressure of water vapour present in the atmosphere}}{\text{s.v.p. of water vapour at the same temperature}}$$

Use your graph to estimate, for a day when the temperature is 23°C and the relative humidity 0.40, (a) the partial pressure of the water vapour in the atmosphere (b) the temperature to which the atmosphere would need to fall for it to become saturated with water vapour.

11-15 It is suggested that the s.v.p. p of ether increases exponentially with temperature T over the range 273 K to 300 K i.e., that $p = p_0 e^{kT}$ where p_0 and k are constants. Plot a straight-line graph which will enable you to test this suggestion and find the temperature interval over which the s.v.p. doubles. (Use the data from the previous question.)

11-16 In a room where the temperature is 20°C a metal surface which is gradually cooled begins to cloud over when its temperature reaches 5°C. Use the information in the previous question to find the relative humidity in the room. Explain your reasoning.

Why do the windows of a warm room, on a cold day, become clouded when there are many people present?

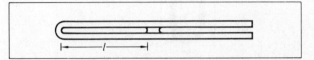

11-17 A capillary tube is sealed at one end, as shown in the figure, and contains a short column of water. Between this and the sealed end is an air column of length l, saturated with water vapour. At 39°C the length l is 98 mm, and at 74°C the length l is 160 mm. If atmospheric pressure is 101.0 kPa, and the s.v.p. of water at 39°C is 7.0 kPa, calculate (a) the partial pressure of the air at 39°C (b) the partial pressure of the air at 74°C (c) the s.v.p. of water at 74°C.

11-18 Explain the difference between evaporation and boiling. Include in your answer a reference to the following points: does either (a) happen only at a particular temperature (b) happen only at the surface of the liquid?

11-19 Some liquid is placed in an open dish and starts to evaporate.
(a) Why does its temperature start to fall?
(b) Why does its temperature stop falling? What is happening then?
(c) If the same volume of liquid had been placed in a wider dish, so that it had a greater surface area, would its temperature have fallen further before becoming steady? Explain.

11-20 Why does a liquid boil at a temperature which depends on the external pressure? Describe an experiment in which use is made of this fact to measure the s.v.p. of a liquid.

11-21 The table shows the variation with temperature θ of the s.v.p. of ether.

θ/°C	− 10	0	10	20	30	40
s.v.p./kPa	14.6	23.9	37.2	57.1	85.1	118.3

Draw a graph of s.v.p. against temperature, and use it to find (a) the normal boiling point of ether (b) the boiling point when the external pressure is 20 kPa.

11-22 A vessel contains air and some ether. Initially the pressure in the vessel is 90.0 kPa and the temperature is 10.0°C. If the temperature is raised to 40.0°C, what is the final pressure in the vessel? Use the data in the previous question, with assume that there is enough ether for the vessel to be always saturated.

Can the ether boil, if the temperature is raised slowly to a sufficiently high value?

11-23 Explain how would you use the apparatus shown in the figure to measure the s.v.p. of a liquid at various temperatures. What does the space above A contain? What difficulties would you meet in trying to obtain an accurate result? For what range of values of s.v.p. is this method most convenient?

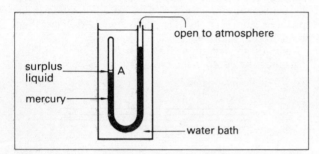

11-24 Refer to the figure for the previous question. If the atmospheric pressure is 101 kPa, and the difference in levels of the mercury is 60 mm, what is the s.v.p. of the liquid at the temperature of the bath? Density of mercury = 13 600 kg m^{-3}.

Is the normal boiling point of the liquid above or below the temperature of the bath?

Describe what would happen if, separately (a) the temperature of the bath was raised (b) some more mercury was poured into the open limb of the manometer.

Heat engines

11-25 The overall efficiency of an Otto cycle internal combustion engine is typically 0.25. Losses in the car transmission system, and in air and road resistance, further reduce this so that the overall efficiency is about 0.10. At what rate is a car heating its surroundings when it is working at a rate of 60 kW?

11-26 An ideal gas ($C_V = \frac{3}{2} R$) is taken round the cycle *abcda* shown in the figure.
(a) Use $pV = nRT$ to calculate nRT for each of the points a, b, c, d.
(b) Use $\Delta U = nC_V \Delta T = \frac{3}{2} nR\Delta T$ to calculate ΔU for each of the processes ab, bc, cd, da.
(c) Use $\Delta W = p\Delta V$ to calculate ΔW for each of the processes ab, bc, cd, da.
(d) Calculate ΔQ for each of the processes ab, bc, cd, da.
(e) For the whole cycle, what is (i) the energy used to heat the surroundings (i.e. wasted) (ii) the energy absorbed from the surroundings by heating (iii) the net work done by the substance (iv) the thermal efficiency of the engine in which this cycle is being used.

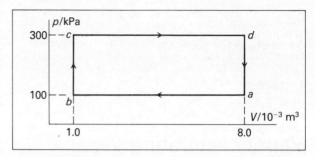

11-27 Sketch a Carnot cycle *abcda* for an ideal gas on a $p–V$ diagram, where ab and cd are the isothermal processes at 300 K and 600 K respectively and bc and da are the adiabatic processes. Suppose that along cd $\Delta Q = + 90$ kJ in each cycle.
(a) What is ΔW for cd?
(b) What is the efficiency of the cycle?
(c) What is (i) ΔQ, (ii) ΔW, for ab?
(d) If $U_a = 50$ kJ, what are (i) U_b, U_c, U_d (ii) ΔQ for bc, da (iii) ΔW for bc, da?

11-28 An ideal Stirling cycle *abcda* consists of the following reversible processes:
ab: isothermal compression from $p = 150$ kPa, $V = 1.0 \times 10^{-3}$ m^3, $T = 300$ K to a volume of 0.50×10^{-3} m^3
bc: increase in pressure at constant volume to $p = 400$ kPa
cd: isothermal expansion to $p = 200$ kPa
da: reduction of pressure at constant volume to the original pressure of 150 kPa.
Find the value of p at point b and the value of T at point c; then calculate the values of p and V at a sufficient number of other points of the processes to enable you to draw an accurate $p–V$ graph for the cycle. Estimate
(a) ΔW, the net work done in the cycle (this is measured by the area enclosed by the cycle)
(b) the energy given to the substance by heating from the surroundings (this is measured by the area between cd and the V-axis).
Calculate the thermal efficiency of the cycle. The Stirling cycle is a Carnot-efficiency cycle, i.e. its thermal efficiency η can be calculated from the equation $\eta = 1 - (T_a/T_d)$. Calculate T_d, and hence η. Does it agree with the result you obtained from your graph?

11-29 The upper temperature in recently-built power stations has been 840 K; older power stations had an upper temperature of about 550 K.
(a) Why has there been this increase?
(b) Calculate the maximum efficiency (i.e. assuming that the turbines use a Carnot-efficiency cycle) if in both cases the lower temperature was 320 K.

(In practice the actual efficiencies may be only three-quarters of these calculated values.)

11-30 The temperature in a freezer is to be kept at $-20.0°C$. When the temperature in the room in which it is kept is $20.0°C$, energy leaks into the freezer at a rate of 15.0 W.
(a) At what average rate must energy be removed from the freezer to keep its temperature constant?

(b) Assume that the freezer uses a Carnot-efficiency cycle and use $Q_c/Q_h = T_c/T_h$ to calculate the energy which must be delivered to the room.

(c) At what rate must the freezer motor work?

(d) Estimate the rate of leakage of energy into the freezer when it is placed in an unheated garage where the temperature is 4.0°C.

(f) If in practice the freezer (when in surroundings of 20°C) uses about 1 kW h of energy a day, what fraction is its efficiency of the efficiency of a Carnot-cycle?

11-31 A heat pump delivers energy to a house at a rate of 11.0 kW: the pump requires a power of 4.5 kW to drive it.

(a) At what rate is energy being obtained from the ground?

(b) By what factor is the heat pump cheaper to run than an electrical resistance heater which is required to deliver the same power to the house? (This is the coefficient of performance η_r of the heat pump.)

(c) The ground temperature was 3°C, and the water in the radiators in the house was at a temperature of 72°C. Calculate the value of η_r for a Carnot-efficiency heat pump working between these temperatures.

The second law of thermodynamics

11-32 The figure shows diagrams which represent four different types of heat engine, i.e. systems operating between a cold and a hot energy reservoir. State whether any of them would contravene the second law of thermodynamics. For the others give a practical example of that type.

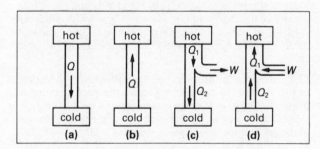

11-33 Imagine a Carnot-efficiency engine working with its cold reservoir at the triple point of water, 273.16 K. The temperature of the hot reservoir is adjusted until, by experiment, it is found that it is taking in 373.16 J of energy from the hot reservoir, and rejecting 273.16 J of energy to the cold reservoir. What is the temperature of the hot reservoir? (Use $Q_h/Q_c = T_h/T_c$ for any Carnot-efficiency cycle.)

What is the 'thermometer' which has been used to make this measurement? [This is a measurement of temperature on the *thermodynamic scale*; since $Q_h/Q_c = T_h/T_c$ implies that T_h measured on the ideal gas scale is also 373.16 K this shows that the ideal gas scale and the thermodynamic scale are identical. Further, since the thermodynamic scale does not depend on the properties of any particular substance, the ideal gas scale is therefore also independent of the properties of a particular substance.]

Entropy

11-34 Calculate the change of entropy when (a) the temperature of 2.0 kg of water rises from (i) 299.5 K to 300.5 K (ii) 369.5 K to 370.5 K (b) 0.20 kg of ice melts at 273 K (c) 0.40 kg of water evaporates at 373 K.

11-35 In the following pairs of processes, which body has the greater change of entropy:

(a) 1.0 kg of water raised from 10°C to 11°C or (b) 2.0 kg of water raised from 10°C to 11°C?

(c) 1.0 kg of water raised from 10°C to 11°C or (d) 1.0 kg of copper raised from 10°C to 11°C?

(e) 1.0 kg of ice melted at 0°C or (f) 1.0 kg of water evaporated at 100°C?

11-36 When energy ΔQ is given to a body at temperature T its change of entropy ΔS is given by $\Delta S = \Delta Q/T$. But $\Delta Q = mc\Delta T$, (where m is the mass of substance and c its s.h.c.) so $\Delta S = mc\Delta T/T$, so if a graph is plotted with $1/T$ on the y-axis and T on the x-axis, the area beneath the graph and the T-axis will give the total change of entropy between any two temperatures, when multiplied by mc. Plot values of $1/T$ against T for values of T from 273 K to 373 K, at intervals of 20 K.

(a) Does the change in entropy of the body for a 1.0 K rise in temperature increase or decrease as the temperature rises?

(b) Estimate the change in entropy for (i) 2.00 kg of water heated from 0°C to 50°C (ii) 2.00 kg of water heated from 50°C to 100°C.

11-37 An ideal gas is reversibly compressed so that its volume is halved (i) isothermally (ii) adiabatically. State whether the following quantities are positive, negative or zero for each of the processes: (a) ΔQ (b) ΔU (c) ΔW (d) ΔS.

11-38 A beaker containing 100 g of water is at 0°C is placed on top of a reservoir whose temperature is 100°C: the reservoir is so large that its temperature does not change perceptibly when it warms the water in the beaker from 0°C to 100°C.

(a) What is the change of entropy of the reservoir?

(b) The change of entropy of the water cannot easily be calculated, since the temperature at which the energy is absorbed changes—but is it numerically less than or greater than the change of entropy of the reservoir? You can tell by considering the range of temperature over which the water gains energy.

(c) Is there a net gain or loss of entropy for the whole system?

11-39 If the Carnot cycle of question **11-27** is drawn on a T–S diagram, with temperature T on the y-axis and entropy S on the x-axis, it is represented by a rectangle, with b the point nearest the origin, and the sense of the cycle being clockwise. Sketch the cycle and explain why (a) bc and da

are constant-entropy processes (b) the entropy is less at b than at a (c) the entropy increases along cd, and decreases along ab.

(d) What is the entropy change for (i) cd (ii) ab?

11-40 Molecules of the gas N_2O_4 dissociate to form two molecules of the gas NO_2. At any temperature the energy needed to break the N—N bond is 57.0 kJ mol^{-1}. The entropies of 1.00 mol of N_2O_4 and NO_2 are 304 J K^{-1} and 240 J K^{-1} respectively. Calculate, for 2.00 mol of N_2O_4 (a)

its change of entropy when it dissociates (b) the energy the surroundings must supply to it to dissociate it (c) the change of entropy of the surroundings if this energy is supplied at (i) 314 K (ii) 324 K (iii) 334 K (d) the *total* change of entropy (i.e. of gas and surroundings) for each of the above three temperatures.

What is the significance of the temperature 324 K?

11-41 If entropy always increases, how does an egg become a chicken?

12 Centripetal forces and gravitation

Data $g = 9.81$ N kg^{-1}
$G = 6.67 \times 10^{-11}$ N m^2 kg^{-2}
mass of Earth $= 5.97 \times 10^{24}$ kg
radius of Earth $= 6.37 \times 10^6$ m
mass of Moon $= 7.34 \times 10^{22}$ kg
radius of Moon $= 1.74 \times 10^6$ m

Describing circular motion

12-1 What is the angular velocity of (a) a flywheel rotating at 5000 r.p.m. (b) the minute hand of a clock (c) a radius of the Earth which links the centre of the Earth to a point on the Equator?

12-2 A synchronous satellite makes a complete circular orbit once in 24 hours. The radius of the orbit is 4.2×10^7 m. What is (a) its angular velocity (b) its speed?

12-3 A gramophone record rotates at $33\frac{1}{3}$ r.p.m., and has a radius of 0.15 m. What is (a) its angular velocity (b) the speed of a point on its circumference (c) the speed of a point midway between its centre and its circumference?

12-4 A turntable has an angular velocity of 3.0 rad s^{-1}. What is its period of revolution?

12-5 A line drawn on a rotating flywheel appears stationary when illuminated by a stroboscopic lamp whose rate of flashing is 50 Hz. When the rate of flashing is raised to 100 Hz, the line again appears stationary. When the rate of flashing is raised to 200 Hz, two stationary lines are seen, opposite each other. Explain these observations. What is the least possible rate of rotation of the flywheel (a) in rev s^{-1} (b) rad s^{-1}? Give (in rev s^{-1}) one other possible rate of rotation.

12-6 The angular velocity of a motor increases steadily from 220 rad s^{-1} to 280 rad s^{-1} in 30 s. What is (a) its average angular velocity (b) the angle through which it turns in that time (c) its angular acceleration?

12-7 A man pulls on a cord to try to start an outboard motor. Initially at rest, after 1.2 s it is rotating at 1200 r.p.m. Assuming that its speed increases uniformly, what is (a) its new angular velocity (b) its angular acceleration

(c) the angle turned through (d) the number of revolutions made by the motor?

12-8 A bucket is tied to a rope, and the rope is wound round a drum which has its axis horizontal: the radius of he drum is 0.50 m. The drum is initially at rest; when the bucket is released it falls for a distance of 9.0 m and is then moving at 6.0 m s^{-1}. What is (a) the angular velocity of the drum (b) the time taken for the bucket to fall (c) the angular acceleration of the drum?

Centripetal forces

12-9 A particle moving in a circular path with centre O and radius r at constant speed v is at some time at a point P. At a time t later it is at Q, where angle POQ $= \theta$. Draw a diagram to show the particle's velocity at P and at Q and prove that the size of the change of velocity (i.e. new velocity minus old velocity, by vector subtraction) is $2v \sin \frac{1}{2}\theta$. Show also that its direction bisects the angle POQ and is directed towards O.

12-10 Refer to the previous question. What is (a) the length of the arc PQ (b) the time taken for the particle to move from P to Q (c) the average acceleration of the particle?

Calculate to 4 significant figures the average acceleration of the particle for the following values of θ: 60°, 40°, 20°, 10°, 5°. As θ tends to zero, what will the average acceleration tend to?

12-11 A particle moves in a circular path at a constant speed. Draw a diagram of the path, mark the position of the particle at some point on the path, and for that point show the direction of (a) its velocity (b) its acceleration (c) the resultant force on it.

12-12 What is the acceleration of a car which has a constant speed of (a) 20 m s^{-1}, and is moving in a circle of radius 100 m (b) 10 m s^{-1}, and is moving in a circle of radius 100 m?

12-13 A point on the Equator makes one revolution in 24 hours. What is (a) the speed of the point (b) its acceleration?

12-14 The bob of a simple pendulum of length 1.2 m is pulled to one side so that it is at a vertical height of 35 mm above its lowest point and is then released. Calculate (a) its speed at its lowest point (b) its acceleration there. Show the direction of its velocity and acceleration (at its lowest point) on a diagram.

12-15 A cyclist turns a corner at a constant speed of 5.0 m s^{-1} in a circular arc of radius 8.0 m. Draw a free-body diagram to show the forces acting on him. What is (a) his acceleration (b) the resultant force on him and his bicycle, if together they have a mass of 90 kg?

12-16 A car of mass 900 kg is driven over a hump-backed bridge at a speed of 18 m s^{-1}. The road surface of the bridge forms part of a circular arc of radius 50 m.
(a) Draw a free-body diagram for the car when it is at the top of the bridge.
(b) Calculate the push of the road on the car then.
(c) What is the greatest speed at which the car may be driven over the bridge if its wheels are not to lose contact with the road?

12-17 A boy ties a stone of mass m onto the end of a string and whirls the stone in a horizontal circle round his head at a constant speed v.
(a) Why cannot the string be horizontal?
(b) Draw a free-body diagram for the stone, marking on it the only *two* forces acting on it (air resistance can be ignored).
(c) The radius of the circle is r, and the string makes an angle θ with the vertical. Apply the equation $ma = F_{res}$ in (i) the vertical direction, and (ii) the horizontal direction, and hence derive an expression for $\tan\theta$ in terms of v, g and r.

12-18 The figure shows a free-body diagram for a car of mass m on a banked motor race track (P and F represent the *total* perpendicular and frictional push of the track on the car respectively). It moves at speed v in a horizontal circle of radius r.
(a) Apply $ma = F_{res}$ in (i) the horizontal direction (ii) the vertical direction.
(b) If $\theta = 26°$, and $r = 300$ m, what value should v have if the frictional force is to be zero?
(c) If a car drives round the turn at a speed of 30 m s^{-1} will the frictional force act in the direction shown, or in the opposite direction?

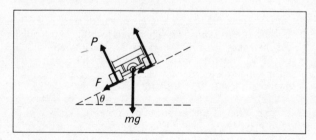

12-19 (a) When you stand on a weighing machine, what force is it which pushes on the weighing machine and makes it record a force? [*Not* your weight.]
(b) Is this force always the same size (while you have the same mass)?

(c) Suggest a situation in which it could be zero.
(d) If it was zero the weighing machine would record that you were 'weightless': would you then have no weight?
(e) Can you directly feel your own weight?

12-20 In which of the following situations would a man *feel* weightless: (a) falling freely near the Earth (b) falling, with an open parachute, from an aircraft (c) on board a space module on its way to the Moon with (i) motors running (ii) motors switched off (d) in a satellite orbiting the Earth?

12-21 Draw a free-body diagram for a satellite in a circular orbit round the Earth, labelling as F the pull of the Earth on the satellite. In a short time Δt its change of displacement is Δs.
(a) Are the following quantities zero: (i) the work done $F\Delta s$ (ii) the impuse $F\Delta t$?
(b) What can you conclude about (i) the kinetic energy of the satellite (ii) the linear momentum of the satellite?

Newton's law of gravitation

12-22 What is the size of the gravitational pull of a sphere of mass 10 kg on a sphere of mass 2.0 kg, when their centres are 200 mm apart? What is the gravitational pull of the 2.0 kg sphere on the 10 kg sphere?

12-23 Calculate (a) the gravitational pull of the Earth on each of the following bodies (b) their acceleration where they are:
(i) the Moon, distance from centre of Earth 3.8×10^8 m
(ii) a geosynchronous satellite, mass 100 kg, distance from centre of Earth 4.2×10^7 m
(iii) a satellite near the Earth, mass 80 kg, distance from centre of Earth 8.0×10^6 m.

12-24 Calculate the gravitational force of attraction between (a) two touching lead spheres of radius 50 mm (b) two touching lead spheres of radius 100 mm. Density of lead = 1.1×10^4 kg m^{-3}.

12-25 When Cavendish (1798) performed his experiment to measure G his large lead spheres had a diameter of 152 mm and a mass of 168 kg and his small lead spheres had a diameter of 51.0 mm and a mass of 6.22 kg. What was the maximum force with which each sphere could have pulled on the other?

12-26 The electrical force of repulsion between two protons (mass 1.7×10^{-27} kg) whose centres are 1.0×10^{-10} m apart is 2.3×10^{-8} N.
(a) What is the gravitational force of attraction between the protons?
(b) What is the ratio (electrical force)/(gravitational force)?

12-27 If the pull of the Moon on a space vehicle is the same size as (but opposite in direction to) the pull of the Earth on the space vehicle when its distance from the centre of the Earth is 9.0 times its distance from the centre of the Moon what is the mass of the Moon? [Hint: write the Earth-Moon distance as $10d$.]

12-28 (a) Use $W = mg$ to find the weight of a mass of 5.00 kg at the surface of the Earth.
(b) Hence use $F = Gm_1m_2/r^2$ to calculate the mass of the Earth.
(c) Deduce the mean density of the material of the Earth.
(d) When the scientists first measured the value of G they said they were 'weighing the Earth'. Explain what they meant.

Gravitational fields

12-29 A mass of 3.0 kg is lifted 1.6 m from the surface of the Earth. What is (a) its change of potential energy (b) the change of potential?

12-30 The figure shows part of the gravitational field near the surface of the Earth. On this scale the field is approximately uniform.
(a) What is the distance apart of the equipotential surfaces?
(b) What is the gravitational force on a mass of 2.0 kg placed at (i) A (ii) B?
(c) What is the gravitational p.e. of a mass of 2.0 kg at (i) A (ii) B?
(d) How much work must be done to move the mass from B to A?
(e) Suppose a stone of mass 2.0 kg has been thrown so that it passes (on a curved path) through A and B: its speed when it passes through A is 3.0 m s^{-1}. What is its speed when it (a) is at B (b) hits the ground?

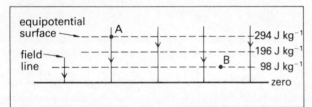

12-31 What is the value of g (the gravitational field strength of the Sun) at the surface of the Sun? Mass of Sun = 2.0×10^{30} kg, radius of Sun = 7.0×10^8 m.

12-32 How far above the surface of the Earth is the gravitational field strength (a) half (b) a quarter, of its value at the surface?

12-33 The figure shows a region near the surface of the Earth.

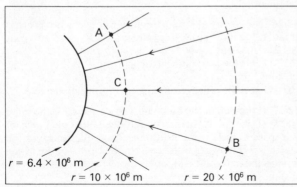

(a) Taking the zero of gravitational potential to be at infinity, calculate the gravitational potential at (i) A (ii) B.
(b) What is the change in g.p.e. of a space capsule of mass 1500 kg when it moves from A to B?
(c) If, later, it returns through B at a speed of 4.0 km s^{-1} how fast will it be moving at C?

12-34 The table gives the speed v of the Apollo 11 spacecraft at different distances r from the centre of the Earth on its journey to the Moon. On this part of the journey it had a constant mass m of 4.4×10^4 kg, and the rocket motors were not being used.

$r/10^6$ m	11.1	26.3	54.4	95.7	170
$v/10^3$ m s^{-1}	8.41	5.37	3.63	2.62	1.80

Plot a graph of the k.e. of the spacecraft against $1/r$, putting k.e./10^{11} J on the y-axis, allowing for values of up to $+18 \times 10^{11}$ J, and also for values down to -18×10^{11} J, for a later part of the question.
(a) What is the intercept of this straight line on the $1/r$ axis? What can you calculate from the value of this intercept?
(b) From your graph estimate the additional k.e. which the spacecraft would have needed if it had been required just to escape from the Earth's gravitational field. [Hint: what would the graph have been like if the k.e. had decreased to zero only at an infinite value of r?]
(c) If, instead, it had been given an additonal 10×10^{10} J, what would have been the speed with which it escaped from the Earth's gravitational field?
(d) Remembering that the sum of the k.e. and g.p.e. of the spacecraft is constant, draw the graph which shows the variation of g.p.e. with $1/r$. [Hint: this is why you need the negative part of the energy axis.] Calculate the slope of this graph: what physical quantity should the slope be equal to? Hence calculate the product Gm_E where m_E is the mass of the Earth.

12-35 Calculate the gravitational pull F of the Earth on a body of mass 10 kg at distances r from the centre of the Earth given by $r/10^6$ m = 6.37 (i.e. at the Earth's surface), 10, 20, 30, 60, 100 and plot a graph of F against r. The area between the graph and the r-axis represents the work done by the force F: estimate the change of gravitational potential energy of the body when it moves from $r = 20 \times 10^6$ m to $r = 80 \times 10^6$ m.

12-36 Use the equation $V_g = Gm_E/r$ (where $Gm_E = 4.0 \times 10^{14}$ N m^2 kg^{-1}) to calculate values of gravitational potential V_g for distances $r/10^6$ m = 6.4, 10, 20, 40, 100, and plot a graph of V_g (on the y-axis) against r (on the x-axis). By drawing tangents to your curve estimate the gravitational field strength g at (a) $r = 30 \times 10^6$ m (b) $r = 60 \times 10^6$ m.

12-37 (a) What is the gravitational potential energy of a body of mass 100 kg at the Earth's surface, taking the zero of potential to be at infinity.
(b) How much k.e. must it be given at the Earth's surface to just escape from the Earth's gravitational field?
(c) What speed would it have then (its escape speed)?
(d) Does this speed depend on the mass of the body? Explain.
(e) Calculate the escape speed for the Moon.

12-38 The temperature of the Moon's surface rises to about 400 K when sunlight falls on it. At this temperature the r.m.s. speed of oxygen molecules is 560 m s^{-1}. Why has the Moon no atmosphere? [Hint: use one of your answers to question **12-37**.]

12-39 A sphere X of radius 0.30 m is placed with its centre at a distance of 2.0 m from the centre of a sphere Y of radius 0.20 m. Both may be imagined to be in a region where the gravitational field caused by other bodies is zero. For the sphere X $Gm_X = 1.0 \times 10^{-7}$ N m^2 kg^{-2} and for sphere Y $Gm_Y = 2.0 \times 10^{-8}$ N m^2 kg^{-2}. Complete the table to show the values of the potential caused by X and Y, separately, at distances r/m = 0.30, 0.50, 0.80, 1.00, 1.20, 1.50, 1.70, 1.80 from the centre of X, and also the total potential $V_X + V_Y$. One line has been completed for you as an example.

r/m	$V_X/10^{-7}$ J kg^{-1}	$V_Y/10^{-7}$ J kg^{-1}	$(V_X + V_Y)/$ 10^{-7} J kg^{-1}
0.30	-3.33	-0.12	-3.45

Draw a graph of $V_X + V_Y$ against r. Estimate the maximum gravitational potential between X and Y. If a particle of mass 2.0 g were to be projected towards Y from the surface of X, what is the minimum amount of k.e. it would need to be able to reach Y?

12-40 Draw a free-body diagram for a man standing on a weighing machine at the Equator. Because he is on the Earth's surface he is moving in a circular path and therefore has a centripetal acceleration. Calculate (a) his acceleration (b) his weight, if his mass is 75.00 kg and the value of g at the Equator is 9.780 N kg^{-1} (c) the push of the weighing machine on him.

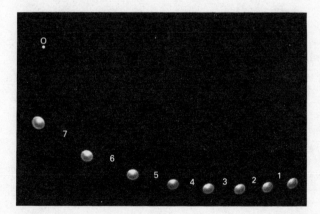

12-41 The figure shows a multiflash photograph of a ball rolling on a curved surface. For each interval 1, 2, 3, etc., measure the separation s of the centres of the images on either side of that interval: why is s^2 proportional to the mean k.e. of the ball in that interval? Also measure, for each interval, the distance r of the centre of the interval from the point O. Is the ball moving faster as it gets nearer O? The k.e. it has gained is (ignoring frictional forces, and any rotational k.e. it may have because it is rolling) equal to the g.p.e. it has lost, so s^2 is proportional to the g.p.e. also. Plot a graph of s^2 against $1/r$. From the shape of your graph deduce the shape of the surface. How is it an analogue of the gravitational field around the Earth?

Satellites

12-42 What is the period of a satellite which is moving a few thousand metres above the Earth's surface? Take $g = 9.8$ N kg^{-1}.

12-43 Calculate the oribtal radius of a synchronous satellite, i.e. one which has a period of 24 hours so that it appears to remain stationary above one point on the Earth's surface. Roughly how many Earth radii is that? Why must that point lie on the Equator?

12-44 The period of Io, the most prominent of Jupiter's satellites, is 1.53×10^5 s, and the radius of its orbit is 4.20×10^8 m. Calculate the mass of Jupiter.

12-45 The Apollo 11 spacecraft was orbiting the Moon at a speed of 1.65 km s^{-1} before the lunar module landed. How far was it above the Moon's surface?

What would be the speed of a satellite moving in orbit just above the surface of the Moon?

12-46 The Apollo 11 space capsule was placed in a parking orbit (around the Earth) of radius 6.56×10^6 m before moving onwards to the Moon.
(a) Calculate the value of the gravitational field strength at this radius.
(b) What was the acceleration of the capsule?
(c) Use $W = mg$ to find the weight of a 70 kg astronaut in the capsule.
(d) What was his acceleration?
(e) Use $ma = F_{res}$ to find the resultant force on him.
(f) Did any force other than his own weight act on him?
(g) What would a weighing machine in the capsule have recorded if he had stood on it?

13 Equilibrium and rotation

Data $g = 9.81 \text{ N kg}^{-1}$

Values of moments of inertia of uniform bodies of mass m (all about an axis through the centre):

thin hoop of radius a, axis perpendicular to plane: ma^2
disc of radius a, axis perpendicular to plane: $\frac{1}{2} ma^2$
solid sphere of radius a: $\frac{2}{5} ma^2$
thin rod of length l, axis perpendicular to rod: $\frac{1}{12} ml^2$

Moments and equilibrium

13-1 The figure shows some forces on a co-ordinate grid. The grid lines are 1.0 m apart. Taking clockwise moments as positive, find the moment of (a) P about (1,1) (b) P about (4,5) (c) Q about O (d) R about O (e) Q about (1,0) (f) S about O.

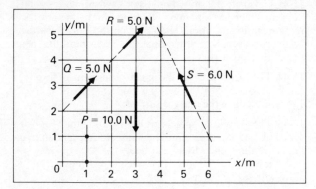

13-2 Two men are walking and carrying a ladder of length 12 m and weight 200 N: it may be considered to be uniform. The first man is 2.0 m from the end of the ladder; the back man holds it at the end. Draw a free-body diagram for the ladder and calculate the force which each man exerts on the ladder.

13-3 A uniform horizontal beam of weight 200 N and length 2.5 m is freely hinged at one end to a wall. A rope is attached to the other end: the other end of the rope is tied to the wall, at a point 1.0 m vertically above the beam. Draw a free-body diagram for the beam, and (a) calculate the tension in the rope (b) calculate the horizontal and vertical resolved parts X and Y of the push of the wall on the beam (c) show on your diagram the direction of the resultant of X and Y.

13-4 A non-uniform tree trunk AB of weight 20 kN and length 10 m lies on the ground. A cable attached at B makes an angle of 40° with the horizontal and pulls with a force of 12.5 kN. The trunk may be considered to be horizontal, with B lifted just clear of the ground, and touching the ground only at A. Draw a free-body diagram for the trunk, and find (a) the position of the centre of gravity (b) the horizontal and vertical resolved parts X and Y of the push of the ground on the tree trunk (c) the resultant of X and Y.

13-5 A uniform ladder of weight 150 N and length 4.00 m leans with its upper end against a frictionless vertical wall. Its lower end rests on the ground, 1.00 m from the foot of the wall. Draw a free-body diagram for the ladder, showing the direction of the push P of the ground on the ladder. Calculate (a) the angle which P makes with the vertical (b) the size of P (c) the push of the wall on the ladder.

13-6 A table lamp has a circular base of diameter 100 mm and is weighted so that when it stands on a table its centre of gravity is 60 mm above the surface of the table. Through what angle can the lamp be tilted before it falls over? [Hint: draw a diagram of the lamp as it is when it is just about to fall over.]

13-7 Part (a) of the figure shows a step-ladder which consists of two hinged parts of equal length; the parts may be considered to be uniform. The left-hand part has a mass of 4.0 kg, and the right-hand part has a mass of 13 kg. The step-ladder is placed on a frictionless floor.
(i) Draw a free-body diagram for the ladder and calculate the size of the two upward pushes of the floor on the ladder.
(ii) Draw a free-body diagram for the right-hand part of the ladder and calculate the tension T in the rope and the vertical and horizontal resolved parts Y and X of the push of the left-hand part of the ladder on the right-hand part.

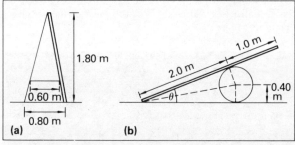

13-8 Part (b) of the figure shows a uniform plank of weight 260 N resting on a circular drum which is fixed to the ground: any frictional force where the plank touches the drum can be neglected. The other end of the plank rests on the ground.
(i) Draw a free-body diagram for the plank.
(ii) Calculate the angle θ.
(iii) Calculate the push of the drum on the plank.
(iv) Calculate the size and direction of the push of the ground on the plank and show the direction on your free-body diagram.

13-9 Part (a) of the figure is a free-body diagram for the spinal column of a person bending over so that his spine makes an angle of 40° with the horizontal. His whole weight is W; the weight of his head and arms is $0.2W$, and the weight of his trunk is $0.4W$. P represents the force which the erector muscles of his spine exert, and R is the push of the sacrum on the lowest spinal disc. With the dimensions shown on the figure, calculate P for a man of weight 700 N.

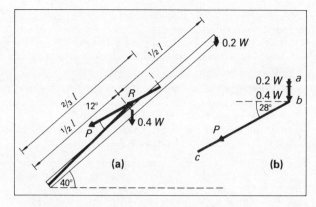

(a)

(b)

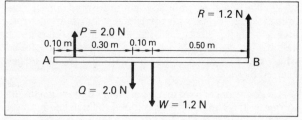

What does *P* become for a woman of weight 600 N who, leaning over in the position shown, picks up a baby of weight 80 N?

13-10 Part (b) of the previous figure shows the start of an attempt to calculate the size of *R* by drawing a scale diagram. The vertical line *ab* represents the weight of the upper part of the body, and the other line *bc* represents *P*. Copy this part of the diagram, taking the man to have a weight of 700 N, and using your previous answer for *P*. Use a scale of 1 mm to 10 N. The resultant of these two forces is represented by a line joining *ac*, so since *R* is the only other force, it must be the same size as the resultant, but opposite in direction. Measure *ac* on your diagram, and hence find the size of *R*.

Couples and torques

13-11 The figure shows a force of 50 N lying along the line *x* = 3 in a co-ordinate framework where the co-ordinates are at one-metre intervals. Its direction is positive. A second force of 50 N lies along the line *x* = 5; its direction is negative. These two forces constitute a couple. What is the moment of this couple about (a) (0,0) (b) (2,0) (c) (4,1) (d) (5,0) (e) (6,3) (f) (− 2,2)?

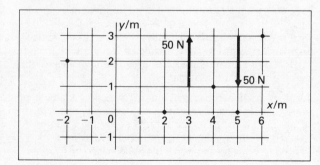

13-12 The figure shows a rod of weight *W* = 1.2 N which has forces *P*, *Q*, *R* exerted on it also.
(a) Taking moments about A, calculate the moment of each of the forces *W*, *P*, *Q*, *R*, taking clockwise as positive. What is the sum of the moments? Is the bar in equilibrium?

(b) What is the moment of the couple caused by (i) *P* and *Q* (ii) *W* and *R*? What is the sum of these moments?
(c) Suppose the couple consisting of the forces *P* and *Q* were moved to the other half of the bar, keeping their separation at 0.30 m, but placing *P* at 0.10 m from the centre. Would the bar still be in equilibrium? Explain.

13-13 Find a paperback book and either measure its mass (using kitchen scales if you are at home) or estimate it. Grip it between your thumb and first finger and lift it, keeping the book upright with its longer edges horizontal. How far from the centre line of the book can you grip it, keeping the longer edges horizontal, using just your thumb and finger? Measure this distance. (If you can manage this with your thumb and finger at the very edge of the book, find a more massive book and try again.) Draw a free-body diagram for the book, marking (a) the pull of the Earth on the book (b) the upward pull of your hand on the book (c) the torque exerted by your hand on the book. Calculate the torque.

13-14 An inn sign consists of a uniform rectangular wooden board, which normally hangs in a vertical plane, hinged along its upper edge. When the hinges are rusty it is found to hang in equilibrium making an angle of 10° with the vertical. Draw a free-body diagram for the sign to show the forces acting on it, and the torque which the hinge exerts on it. The mass of the sign is 9.0 kg, and its height is 1.2 m. Find (a) the upward pull of the hinge on the sign (b) the frictional torque which the hinge exerts on the sign.

13-15 A motor car road test report states that the *single* force needed to turn a steering wheel of diameter 0.36 m is 110 N when the car is stationary, the force being applied at the rim of the wheel. Draw a free-body diagram showing the forces, the torque, and any other force exerted on the steering wheel, and calculate the torque.
Also draw a free-body diagram for the situation where two forces of the same size, acting in opposite directions, are used to turn the steering wheel, acting at opposite ends of a diameter. How large is each of these forces?

13-16 A man pulls open a door fitted with a door closing mechanism. The width of the door is 1.2 m, and the handle is at the edge of the door. Draw a free-body diagram for the door when he is holding it open at an angle of 30° with its original position, assuming that he is pulling at right angles to the original position of the door. If he exerts a force of 9.0 N and the door is then in equilibrium what torque is exerted by the door closer? What force would he have to exert to do the same thing if the handle had been 0.20 m from the line of the hinges?

13-17 The power of a motor is being measured using a band brake. When the motor is rotating at 3000 r.p.m. the readings on the spring balances attached to the two ends of the belt are 140 N and 30 N, and the radius of the pulley round which the belt passes is 0.32 m. Calculate (a) the torque exerted by the motor (b) the power of the motor.

13-18 A motor car road test report states that the maximum torque is 156 Nm at 3400 r.p.m. and the maximum power is 69 kW at 4900 r.p.m. Find (a) the torque at 4900 r.p.m. (b) the power at 3400 r.p.m. Sketch a rough graph of torque against angular speed for this car.

13-19 A cyclist can maintain a rate of working of 200 W. If he pushes the pedals round 5 times in 2.0 s, what average torque does he exert? If he pushes on the pedals at a distance of 0.18 m from the axis of rotation, what is the average force which he exerts?

13-20 The current supplied by a car battery when it is turning the starter motor of a car is 170 A; the p.d. across the battery terminals is then 9.2 V. Assuming that the motor is 80% efficient, calculate the torque exerted by the starter motor on the engine if it rotates it at 700 r.p.m.

13-21 Two meshing gear wheels have 50 and 150 teeth respectively, and radii of 18 mm and 54 mm. A torque of 5.0 Nm is exerted on the shaft of the first gear wheel, so that it rotates, and thus turns the second gear wheel.
(a) What force does the first gear wheel exert on the second gear wheel?
(b) What torque does the first gear wheel exert on the second gear wheel?
(c) Is there here a contradiction of the principle of conservation of energy? Explain.

Moment of inertia

13-22 What is the moment of inertia of a uniform solid cylinder of radius a and length l about its central axis? Explain.

13-23 What is the moment of inertia of (a) a cylindrical shell of radius 0.30 m and length 1.0 m made of sheet metal which has a mass of 20 kg per square metre (b) a cylindrical shell made of the same sheet metal, and of the same length, but with twice the diameter?

13-24 A uniform rod of length l has a moment of inertia about an axis through its centre, and perpendicular to its length, of $\frac{1}{12}ml^2$.
(a) What is the moment of inertia of a rod of length $2l$ and mass m about its midpoint O?
(b) If this rod is now folded at O so that the two halves are together the moment of inertia about O is $\frac{1}{3}ml^2$. Explain.
(c) What is the moment of inertia of a uniform square sheet of side l about an axis along one edge?

13-25 What is the kinetic energy of a steel flywheel which has the shape of a uniform disc of thickness 30 mm and

diameter 0.60 m when it is rotating at (a) 20 rad s^{-1} (b) 40 rad s^{-1}? Density of steel $= 7.7 \times 10^3$ kg m^{-3}. What linear velocity would a lump of steel of this mass have if its k.e. were the same as the flywheel which was rotating at 20 rad s^{-1}?

13-26 What is the kinetic energy which the Earth has because (a) it is rotating about an axis through its poles (b) it moves in an orbit around the Sun? Mass of Earth $= 6.0 \times 10^{24}$ kg, radius of Earth $= 6.4 \times 10^6$ m, mean Earth-Sun distance $= 1.5 \times 10^{11}$ m, one year $= 3.2 \times 10^7$ s.

13-27 For linear motion the equation $Fs = \frac{1}{2}mv^2 - \frac{1}{2}mu^2$ relates the work done by a force to its change of kinetic energy. What similar equation holds for angular motion?
A man pulls steadily for a distance of 0.80 m with a force of 16 N on a rope wound round a pulley. It starts from rest and he finds that his hand is moving at 0.60 m s^{-1} after some time. If the pulley has a radius of 0.35 m, what is (a) the angular velocity of the pulley then (b) its moment of inertia about its central axis?

13-28 A uniform square trap door of side 0.80 m and mass 12 kg is hinged about one edge and is normally horizontal. The edge opposite the hinge is lifted until the trapdoor is vertical. Then the trapdoor is allowed to fall back to its original position. When it reaches the horizontal position what will be (a) its angular velocity (b) the speed of the edge opposite the hinges? [Hint: use the result of part (c) of question **13-24**.]

13-29 A flywheel is fixed to an axle which rotates in bearings which may be assumed to be frictionless. A thread is fixed to the axle of radius 15 mm and wrapped round it several times; the other end is attached to a mass of 50 g. When the mass is allowed to fall the flywheel and axle rotate. The mass took 52.3 s to fall 1.73 m. Find (a) the final speed of the mass (b) the final angular velocity of the flywheel and axle (c) the loss of g.p.e. of the mass (d) the moment of inertia of the flywheel and axle, assuming that the k.e. of the mass is small enough to be neglected. Is this assumption justified?

13-30 Explain what is meant by the angular acceleration of a body.
A pulley has a radius of 60 mm and a moment of intertia of 1.0×10^{-2} kg m^2. A man pulls with a steady force of 30 N on a rope wound round the pulley. What is (a) the pulley's angular acceleration (b) its angular velocity after 2.0 s if it started at rest (c) the angle it has turned through then (d) the number of revolutions it makes?

13-31 A friction brake acts on a flywheel of moment of inertia 2.0 kg m^2 and slows it down from an angular velocity of 500 rad s^{-1} to an angular velocity of 340 rad s^{-1} in 4.0 s.
(a) What is the angular acceleration of the flywheel?
(b) What is the torque which the brake exerts on it?

13-32 A rolling disc has kinetic energy because it is moving linearly *and* because it is rotating. If two discs A and B have the same mass and the same radium but A has the larger moment of inertia, and both are allowed to roll down the same slope, starting from rest, which will reach the bottom of the slope first?

Angular momentum

13-33 A bicycle wheel (of moment of inertia 0.20 kg m^2) on an upturned bicycle is rotating at 25 rad s^{-1}. If it stops in 90 s what is the frictional torque which the bearings exert on the wheel?

13-34 What is the angular momentum which the Earth has because it is (a) rotating about an axis through its poles (b) moving in orbit around the Sun? Use the data from question **13-26**.

13-35 A diver jumps off a diving board with her body straight, and gives herself an angular velocity of 5.0 rad s^{-1}. While in the air she brings her knees up to her chin and in this position her moment of inertia is reduced to one-third of its original value. What is (a) her new angular velocity (b) the ratio of her rotational k.e. now to its original value?

She untucks her body when entering the water to try to enter it cleanly with her body in a vertical line. Why can she not achieve this completely?

13-36 A turntable in a laboratory has a moment of inertia of 0.025 kg m^2 and is rotating at 10 rad s^{-1}. A disc is dropped onto the turntable, coaxially with the turntable, and the angular velocity of the turntable with the disc on it falls to 6.8 rad s^{-1}. What is (a) the moment of inertia of the disc about an axis through its centre (b) the amount of mechanical energy converted to internal energy? Explain how this conversion is caused.

13-37 The only significant force acting on the planets of the solar system is the gravitational pull of the Sun. Explain why this implies that the angular momentum, about an axis through the Sun, of a planet is constant.

At its greatest distance from the Sun, 1.522×10^{11} m, the Earth is moving at a speed of 2.93×10^4 m s^{-1}. What is its speed at its least distance from the Sun, 1.471×10^{11} m?

13-38 A man of mass 75 kg stands at the edge of a circular turntable of radius 2.8 m and moment of inertia 90 kg m^2. If he walks round the edge of the turntable at a speed of 0.60 m s^{-1} (clockwise as seen from above) relative to the ground, what is the angular velocity of the turntable while he is walking, and what is its direction? What is his speed relative to the turntable?

14 Storing electric charge

Data $g = 9.81$ N kg^{-1}

Capacitors

14-1 In the arrangement shown in the figure the meter is a sensitive light-spot type with its zero set in the centre of its scale. The two-way switch is initially in position Y. Describe what you would observe, and account for it in terms of the movement of charge in the circuit, (a) when the switch is moved to position X (b) when the switch is then lifted off contact X and held mid-way between X and Y without touching either (c) when the switch is moved again to position Y.

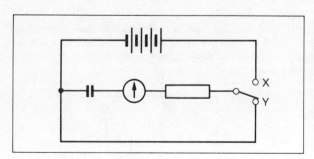

14-2 In the previous question a c.r.o. may be used to observe what is happening in the circuit. Describe what traces you would expect to see on the screen in each of the three stages (a) (b) and (c) of that question, (i) if the c.r.o. is joined across the resistor (ii) if the c.r.o. is joined across the capacitor.

14-3 A certain type of microphone consists of a strip of metal foil separated from a metal plate by a narrow air gap and insulated from it. The plate and the foil are connected to a suitable d.c. supply. Sound waves arriving at the foil cause it to vibrate to and fro varying the width of the air gap. Describe what changes you would expect to occur in the charges on the foil and on the plate as the foil vibrates. At what point in the vibration of the foil would (a) the charge on the foil be a maximum (b) the current in the connecting leads be a maximum?

Capacitance

14-4 A capacitor of capacitance 10 μF is connected to a battery of e.m.f. 12 V. What are the charges on its plates?

14-5 What potential difference must be applied between the plates of a 100 pF capacitor for the charges on them to be ± 0.025 μC?

14-6 In the apparatus shown in question **14-1** the calibration constant of the meter (used as a ballistic galvanometer) is 1.3×10^{-8} C div^{-1}. The battery is of e.m.f. 6.0 V. When the switch is moved from position Y to position X, the maximum swing of the light-spot is found to be 22 divisions. What is the capacitance of the capacitor?

14-7 What charge is carried on each of the plates of a 100 μF capacitor when there is a potential difference between the plates of 200 V? If this capacitor is charged through a large resistance in 4.0 s, what is the average charging current? Will the charging current at the start of this time be greater or less than the average current?

14-8 A 33 μF capacitor is charged and then insulated. During the next minute the potential difference between the plates falls by 4.0 V. What is the average leakage current between the plates?

14-9 A steady charging current of 50 μA is supplied to the plates of a capacitor and causes the potential difference between them to rise from 0 to 5.0 V in 20 s. What is the capacitance?

14-10 A capacitor of capacitance 0.47 μF is connected across the Y-input terminals of a c.r.o. The horizontal (time-base) control is set at 100 ms div^{-1}, and the Y-input control at 5.0 V div^{-1}. A battery is then connected across the capacitor for a short period of time t, and the trace shown in the figure is observed.
(a) Estimate the period of time t.
(b) Estimate the change in p.d. across the capacitor.
(c) What is the rate of change of p.d. (dV/dt)?
(d) What can you say about the charging current?

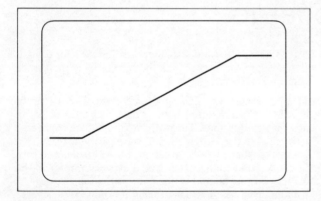

14-11 A capacitor is joined in series with a resistor, a low resistance microammeter, a switch and a 12 V battery, as shown in the figure.
(a) When the switch is first closed, what are (i) the p.d. across the capacitor. (ii) the p.d. across the resistor (iii) the current (iv) the charge on each of the capacitor plates?
(b) Calculate also the same quantities when the switch has

been closed for some time so that steady conditions have been reached.

If the capacitor continued to charge at its initial rate in this arrangement, how long would it take to become fully charged?

14-12 In the previous question, at a certain instant after the switch has been closed the microammeter reads 9.5 μA. Calculate for this instant (a) the p.d. across the resistor (b) the p.d. across the capacitor (c) the charge on each of the capacitor plates.

14-13 In question **14-11** after several seconds of charging the switch is opened again. The capacitor is then discharged through a ballistic galvanometer, and it is found that the charge recorded is 9.5×10^{-5} C. Calculate for the instant just before the switch was opened (a) the p.d. across the capacitor (b) the p.d. across the resistor (c) the micro-ammeter reading.

14-14 A capacitor of capacitance 180 pF is charged to a potential difference of 12 V and then discharged through a sensitive meter, and this sequence of operations is repeated by means of a reed-switch 250 times per second. What is the average current in the meter?

14-15 A vibrating reed is used to connect a capacitor alternately to a battery and to a meter. In this way the capacitor is fully charged by the battery and fully discharged through the meter 50 times per second. Give a circuit diagram showing how this arrangement is set up. Explain the purpose of the diode joined in series with the reed-switch coil in this arrangement.

If the e.m.f. of the battery is 12 V and the meter records an average current of 2.4 mA, what is the capacitance of the capacitor?

14-16 Plot a graph showing how the charge Q in a 1.5 μF capacitor varies with the potential difference V across it for values of p.d. from 0 up to 30 V.

On the same axes draw a further two lines to show the relation between Q and V for a pair of such capacitors (a) in parallel (b) in series.

14-17 Three capacitors of capacitances 2.0 μF, 3.0 μF and 6.0 μF are joined (a) in parallel (b) in series. What is the combined capacitance in each case?

14-18 In the arrangement shown the battery has an e.m.f. of 9.0 V, and the capacitors A, B and C have capacitances of 3.0 μF, 1.5 μF and 4.5 μF respectively. Work out the charge stored in each capacitor and the p.d. between its plates.

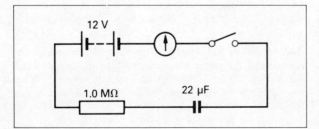

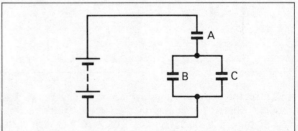

14-19 If you have three capacitors of capacitances 3.0 μF, 6.0 μF and 8.0 μF, how could you produce a combination of capacitance 10 μF?

14-20 In a certain piece of equipment you need to replace a 1 μF capacitor which is used with potential differences up to 500 V. All you have available is a stock of 1 μF capacitors tested to work up to 200 V. What combination of these would you use?

14-21 A capacitor of capacitance 1.5 μF is joined across an alternating supply of peak p.d. 15 V and of frequency 50 Hz. What is the peak value of the current in the capacitor?

14-22 A capacitor is connected in series with an a.c. ammeter to a sinusoidal a.c. supply of frequency 10 kHz. An a.c. voltmeter joined across the supply reads 1.5 V, and the ammeter reads 0.21 A. Calculate the capacitance of the capacitor.

Electrometers

14-23 An electrometer of the electronic type is designed to measure p.d.s. of up to 1.0 V, and has an input resistance of 1.0×10^{13} Ω. Explain how such an instrument may be adapted to measure (a) currents up to 10 pA (b) charges up to 1.0 nC.

14-24 An electrometer as described in the previous question is designed to be operated with its output connected to a moving-coil meter of resistance 500 Ω that gives full-scale deflection for a current of 1.0 mA. The unit is described as a form of amplifier. Explain what the properties of this amplifier must be to achieve the desired result. By what figure must its input p.d. and current be amplified? By what factor is its output power greater than the power supplied to its input?

14-25 A capacitor of unknown capacitance is charged by conecting it momentarily across a 30 V d.c. supply. It is then touched across the input terminals of an electrometer which is being operated with an input capacitance between its terminals of 0.010 μF. The instrument reads 0.25 V. Calculate (a) the charge delivered to the electrometer (with its fixed capacitor) (b) the unknown capcitance (c) the charge in the unknown capacitor when the p.d. across it is 0.25 V (d) the percentage of the charge in the unknown capacitor that remains in it after it has been touched across the electrometer terminals.

14-26 A capacitance of 1.00×10^{-8} F is connected between the terminals of an electrometer in order to perform the following experiment. A capacitor of unknown capacitance is charged from a 25 V battery and then joined across the terminals of the electrometer; this process is repeated 10 times without discharging the electrometer. The electrometer reading is then found to be 0.28 V. Calculate the unknown capacitance.

What percentage of the charge on the unknown capacitor remains in it after it has been connected to the electrometer on the tenth occasion?

14-27 A capacitor of capacitance 1000 pF is connected between the terminals of an electrometer. Another capacitor is charged by joining it momentarily across a small battery; it is then joined across the terminals of the electrometer, and the instrument reads 0.30 V. When the battery itself is joined directly across the electrometer the reading is 0.80 V. What is the capacitance of the second capacitor? What percentage of the charge on the second capacitor is delivered to the electrometer (and its fixed capacitor) in this experiment?

14-28 A leaf electrometer has a capacitance of 15 pF. When a potential difference of 800 V is applied between cap and case the leaf deflects through 8 divisions of the scale. In this position, owing to imperfect insulation, the deflection is found to decrease slowly at the rate of one division in five minutes. Assuming that the deflection is proportional to the applied p.d., estimate (a) the leakage current (b) the resistance of the insulation.

14-29 The graph in the figure shows how the deflection θ of a leaf electrometer varies with the p.d. V between cap and case. The capacitance of the instrument is 12 pF.
(a) What is the p.d. indicated by a deflection (i) of 60° (ii) of 30°?
(b) If it takes 12 hours for the deflection of the instrument to fall from 60° to 30°, estimate the average leakage current during this time.
(c) Estimate the resistance of the insulation between cap and case.

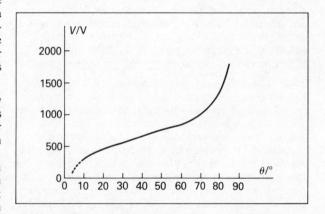

14-30 Refer to the graph in the previous question showing how the deflection θ of a leaf electrometer varies with the p.d. V applied across it.
(a) Is there any range of p.d.s. over which the variation could be described as linear?
(b) Is there any range of p.d.s. over which θ could be described as directly proportional to V?
(c) If the uncertainty in estimating θ is in practice about 5°, what is the uncertainty in the p.d. recorded by the instrument (i) at 500 V (ii) at 1500 V?
(d) If an alternating p.d. of 1000 V (r.m.s.) is connected across the electrometer, what deflection would you expect?

Charging and discharging

14-31 A capacitor of capacitance 22 μF is charged by connecting it to a 400 V supply, and is then discharged.
(a) Calculate the energy converted during the discharge.
(b) If the discharge takes 10 μs, what is the average power of the discharge?

14-32 To what p.d. must a 1500 μF capacitor be charged to enable it, when it is discharged through a small electric motor, to lift a 10 g mass through a vertical height of 1.0 m? (Assume that 10% of the electrical energy is converted to gravitational potential energy in this process.)

14-33 A capacitor of capacitance 16 μF is connected for a short time across a 150 V d.c. supply. What is the charge on the capacitor plates, and what is the energy stored?

A second (initially uncharged) capacitor of capacitance 8.0 μF is now joined in parallel with the first. Calculate the new potential difference across the capacitors. How much energy is now stored in the system? How do you account for the change of energy?

14-34 An 8.0 μF capacitor is charged by joining it to a 500 V supply through a resistor.
(a) How much electrical energy is taken from the supply?
(b) How much electrical energy is stored in the capacitor?
(c) How do you account for the difference between these two amounts?

14-35 After the capacitor in the previous question has been charged to a p.d. of 500 V it is disconnected from the supply and joined across a neon bulb which conducts only so long as the potential difference across it is more than 100 V.
(a) What electric charge passes through the bulb?
(b) How much electrical energy does the capacitor lose?

14-36 Two capacitors, one of capacitance 8.0 μF charged to a potential difference of 50 V, and the other of capacitance 4.0 μF charged to a potential difference of 200 V, are joined in parallel (positive terminals together). Calculate
(a) the charge on each capacitor before they are joined together (b) the potential difference across them after they are joined together (c) the charge that flows from one capacitor to the other (and indicate on a diagram exactly from where to where this charge flows) (d) the decrease in the energy stored as the result of joining the capacitors together.

14-37 A steady p.d. of 200 V is maintained across a combination of two capacitors in series, one of capacitance 2.0 μF and the other of capacitance 0.50 μF. Calculate (a) the combined capacitance (b) the charge stored in each capacitor (c) the potential difference across each capacitor (d) the energy stored in each capacitor.

14-38 Two capacitors of capacitance 5.0 μF and 3.0 μF are joined in series. What is their combined capacitance? What capacitor joined in parallel with this combination will produce a combined capacitance of 4.0 μF?

What energy will be stored in this system when it is connected to a d.c. supply of 12 V?

14-39 A capacitor of capacitance 6000 μF is charged to a potential difference of 50 V. It is then discharged through a compact tangle of copper wire of total mass 2.5 g. If the s.h.c. of copper is 380 J K^{-1} kg^{-1}, calculate the rise in temperature produced.

14-40 In the arrangement shown in the figure the three 15 μF capacitors are initially joined in parallel across the 50 V supply (by having all the switches in the A position). The five switches are then all simultaneously moved to the B position, thus joining the capacitors in series.
(a) What is now the potential of the point X?
(b) If X is joined to Earth through a 10 MΩ resistor, what is the initial current in this?
(c) How much internal energy is produced in the resistor?
(d) Is the middle capacitor discharged in this process? If so, where have the charges on its plates gone?

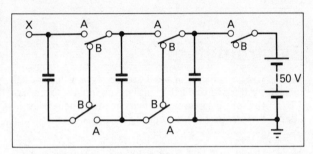

14-41 A capacitor of capacitance 10 μF is charged to a potential difference of 20 V and then isolated. If its leakage resistance is 10 MΩ, how long will it take for the potential difference across it to fall to 10 V?

14-42 A capacitor is joined across an electrometer and charged to a potential difference of 1.00 V. The potential difference V is then measured at 20 s intervals, as tabulated below. When time $t = 30$ s, a resistance R of 1.5 MΩ is joined across the capacitor.

t/s	0	20	40	60	80	100	120
V/V	1.00	1.00	0.81	0.54	0.35	0.23	0.15

(a) What is the current in the resistor at $t = 30$ s?
(b) Plot a graph of V against t, and measure the rate of decrease of V immediately after $t = 30$ s.
(c) Hence calculate the capacitance C of the capacitor.
(d) Plot also a graph of $\ln(V/V)$ against t to demonstrate the exponential fall of p.d., and measure the time constant (RC) of the decay process.
(e) From the time constant calculate the capacitance C. Explain which method of finding C gives the best estimate, and why.

14-43 A capacitor is joined in series with a 150 kΩ resistor, a microammeter and a switch to a d.c. supply. At time $t = 10$ s, the switch is closed, and readings of the current are taken until $t = 80$ s. A graph of I against t is shown in the figure.
(a) Calculate the p.d. of the supply.
(b) At what moment does the p.d. across the resistor fall to exactly half its initial value?

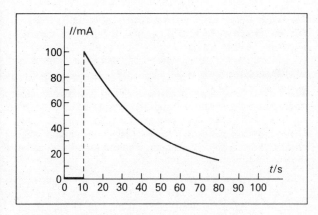

(c) The area under the graph represents the charge that flows through the resistor. By estimating the area of strips of the graph representing successive time intervals of 10 s, calculate and tabulate the total charge Q in the capacitor at 10 s intervals from $t = 10$ s to $t = 80$ s.
(d) Plot a graph of Q against t.
(e) What is the total charge in the capacitor at the moment given by (b) above?
(f) What is the p.d. across the capacitor at this moment?
(g) Calculate the capacitance of the capacitor.

14-44 A poor quality 100 μF capacitor is charged and connected to an electrometer (of very high resistance). It is found that the p.d. across the capacitor falls to half its initial value in 300 s. What is (a) the time constant of the discharge process (b) the resistance of the insulation of the capacitor?

14-45 A reed-switch kept vibrating at a rate of 50 s^{-1} is used to connect a capacitor of capacitance 1.0 nF alternately to a 10 V supply and a 1.0 kΩ resistor. The time during which the capacitor is connected neither to the supply nor to the resistor can be taken as negligible.
(a) If an oscilloscope is joined across the resistor, indicate the type of trace you would expect to observe.
(b) What is the pead p.d. across the resistor?
(c) What is the peak current in the resistor?
(d) What is the average current in the resistor?

14-46 A 400 μF capacitor is charged to a p.d. of 100 V. It is then joined across a resistor of resistance 250 kΩ, which is of heat capacity 0.050 J K^{-1} and is initially at a temperature of 290 K. Assuming that the resistor is thermally isolated, calculate (a) the intial current (b) the initial rate of conversion of electrical energy to internal energy (c) the initial rate of rise of temperature (d) the temperature of the resistor after it has been connected across the capacitor for 50 s.

15 Electric fields

Data $g = 9.81$ N kg^{-1}
$e = 1.60 \times 10^{-19}$ C
$\epsilon_0 = 8.85 \times 10^{-12}$ F m^{-1}
rest mass of an electron $m_e = 9.11 \times 10^{-31}$ kg

Electrical forces

15-1 An experimenter arranges a pair of metal plates in parallel vertical planes, insulates them from their surroundings and connects them to the terminals of a high-voltage supply. To study the properties of the electric field between the plates he performs the following tests. Describe and explain what you would expect to observe in each case.
(a) He pulls a plastic pen from his pocket, rubs it on his sleeve, and then ties a fine thread to its centre; this he uses to suspend the pen in the space near the plates with one end in the gap between them.
(b) He takes a bunch of fine conducting fibres and scatters them in the air above the plates.
(c) He makes a dry mixture of powdered red lead and yellow sulphur and puffs it from a metal nozzle into the gap between the plates. (It is known that red lead gains a positive frictional charge, and yellow sulphur a negative one, in contact with metal.)

15-2 Some small fragments of paper are scattered on a wooden bench top. When a polythene strip that has been rubbed with a woollen duster is moved across the bench top a few centimetres above it the paper fragments leap up and stick to it; a few seconds later some of the paper fragments suddenly leave the surface of the polythene strip, dive down to the bench top and quickly up again to the strip. Explain what is going on in terms of the electric charges involved.

Another experimenter does the same test, but has left a small gas burner alight at the back of his bench. He observes no movement of the paper fragments at all. Explain why this is so.

(You can try both these experiments on a table at home. A plastic ball point pen can take the place of the polythene strip.)

15-3 To illustrate the nature of an electric current in a conducting medium (such as an electrolyte) between a pair of metal plates a teacher constructs the model in figure (a) below. A plastic cylinder is set up vertically and closed at the bottom by a metal sheet. A circular metal disc is attached to a rod and fixed inside the cylinder so that there is a gap of a few centimetres between the base plate and the disc. A handful of very small metallised polystyrene balls is placed in the gap; and the base plate and disc are then connected across a variable high-voltage supply. Explain what you would expect to observe as the p.d. of the supply is steadily increased.

Discuss how far this model adequately illustrates the process of conduction in an electrolyte, and point out the respects in which it could be misleading.

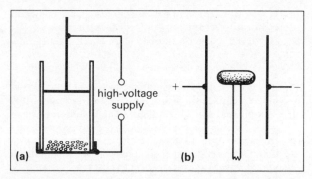

(a)   high-voltage supply   +   −   **(b)**

15-4 A pair of metal plates is arranged as in figure (b), and connected across a high-voltage supply. In the gap between them a small conducting object is fixed on an insulating stand. Copy the diagram and mark in on it the signs of the charges you would expect to find on the various conducting surfaces; sketch in the lines of force of the electric field. List the electrical forces that act (a) on each of the plates (b) on the conducting object in the middle.

An experimenter now takes a short length (5 mm) of copper wire and suspends this at its centre from a fine nylon thread. Explain what you would expect to observe when this is lowered carefully into the gap between the plates. What will happen if it touches either of them?

15-5 (a) What electrical force acts on a small polystyrene ball carrying a charge of 5.0 nC in a horizontal electric field of strength 0.50 MV m^{-1}?
(b) If this polystyrene ball is of mass 0.30 g, what gravitational force acts on it in the Earth's gravitational field?
(c) If the polystyrene ball is suspended by a fine insulating thread in both the above fields, draw a free-body diagram for the ball, and calculate the angle to the vertical at which the thread will come to rest in the above two fields.

15-6 By considering the defining equations of the quantities involved show that 1 V m^{-1} = 1 N C^{-1}.

15-7 A charged polystyrene ball of mass 0.14 g is suspended by a nylon thread from a fine glass spring. In the absence of any electric field the spring extends by 30 mm. The polystyrene ball is then placed in an electric field that acts vertically upwards, of strength 2.0×10^5 V m^{-1}, and the spring extends by a further 6.0 mm. What is the electric charge on the polystyrene ball?

15-8 What radius of water drop carrying a surplus charge of 1 electron would remain stationary under the combined action of the Earth's electric and gravitational fields? Show on a diagram the direction of the lines of force of the electric field near the ground. (Take the Earth's electric field strength as 300 V m^{-1}, and the density of water as 1000 kg m^{-3}.)

Electrical potential

15-9 A small metal object in contact with an earthed plate carries a charge of 8.0×10^{-9} C. How much work is done on it by the electric field if it is moved to another plate at a potential of -3.5 kV?

15-10 An electron is emitted with negligible energy from an earthed electrode in a vacuum tube. It is then accelerated towards another electrode maintained at a potential of 60 V. Calculate (a) the kinetic energy gained by the electron (b) the speed it reaches.

15-11 Two parallel metal plates are fixed 20 mm apart and are maintained at a potential difference of 1000 V.
(a) What is the potential gradient in the gap?
(b) What force would act on a particle in the gap carrying a charge of 2.0×10^{-11} C when its distance from the negative plate is (i) 10 mm (ii) 2.0 mm?

15-12 A vacuum tube contains two plane parallel electrodes 7.5 mm apart. If a p.d. of 150 V is maintained between them, what is (a) the electric field strength in the gap (b) the force acting on an electron in the gap?
(c) If an electron is emitted at negligible speed from the negative electrode, how long does it take to cross the gap?

15-13 Four flat metal plates A, B, C and D are arranged parallel to one another, as in figure (a). The plates are each 1.0 mm thick, and the distance between adjacent plates is 25 mm. The two outer plates A and D are earthed, while B and C are maintained at potentials of $+150$ V and $+450$ V respectively. Draw a graph to show how the potential varies along a line perpendicular to the plates from plate A through to plate D. What is the electric field strength at different positions along this line, both between the plates and inside the metal (you should give both the direction and magnitude of the field)?

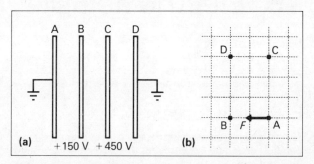

     A   B   C   D   D   C

                         B   F   A

(a)   +150 V   +450 V   **(b)**

15-14 Figure (b) shows a small charged particle at a point A in a *uniform* electric field. The particle experiences an electrical force F as shown of 5.00×10^{-7} N. The grid lines in the figure are at intervals of 10.0 mm. Calculate the work done by the electrical force if the particle is moved (a) from A to B (b) from A to C (c) from A to D.

If the particle carries a charge of 2.50×10^{-11} C, and the point A is at a potential of 200 V, what are the potentials of B, C and D? Which of the grid lines in the figure coincide with lines of force and which with equipotential surfaces?

If the field described in this question is in fact produced by a pair of flat metal plates, one of which is earthed and the other of which is at a potential of 1000 V, copy the diagram and draw in on it the positions of the two plates.

15-15 What is the energy (in MeV) gained by (a) an electron (b) a proton (c) a helium nucleus on being accelerated through a potential difference of 10 MV?

15-16 A proton can be taken as having a mass m and a charge +e, while an α-particle has a mass 4 m and a charge +2e. When these particles are accelerated from rest through the same potential difference, what is (a) the ratio of the energies they gain from the electric field (b) the ratio of the speeds they attain?

Millikan's experiment

15-17 In Millikan's oil drop experiment we observe the drops to move with constant *speed*. By drawing free-body diagrams for an oil drop both when the electric field is switched on and when there is no electric field, explain why the speed rather than the acceleration of the oil drop is constant.

15-18 Describe how you would measure the charge carried by one electron. Explain how you would use your measurements to convince a sceptical observer that the charge you have measured is the smallest charge that can be carried by any particle.

15-19 A charged oil drop of mass 2.0×10^{-15} kg is observed to remain stationary in the space between two horizontal metal plates when the p.d. between them is 245 V and their separation is 8.0 mm. What is the charge on the drop?

15-20 What potential difference would you need to maintain between two horizontal metal plates 6.00 mm apart so that a particle of mass 4.00×10^{-15} kg with three surplus electrons attached to it would remain in equilibrium between them? Which plate would be the positive one?

15-21 Refer to the previous question. When the potential difference between the plates is 480 V, the particle moves slowly downwards in the space between them. When it has fallen 4.00 mm, what is the work done (a) by the gravitational force acting on it (b) by the electrical force acting on it? (c) What is the change in its potential energy (gravitational + electrical)? Explain fully the energy conversions involved in this movement.

Permittivity

15-22 The fair weather electric field strength at the Earth's surface is 300 V m^{-1} directed vertically downwards. What is the density of charge on the Earth's surface and of what sign is it? If the surface area of the British Isles is 3.0×10^{11} m^2, what is the total surface charge carried in fair weather on this part of the Earth's surface?

15-23 The surface density of charge on a certain region of a metal object is 1.6×10^{-5} C m^{-2}. What is the strength of the electric field close to this surface?

15-24 A capacitor consists of two parallel metal plates 12 mm apart in air, each of area 0.040 m^2. If the potential difference between the plates is 4.0 kV, calculate (a) the electric field strength in the gap (b) the surface charge density on each plate (c) the total charge on each of the plates.

(d) What would the total charge on the plates be if the gap between them was filled with an oil of relative permittivity 2.5?

15-25 A capacitor is made by coating a strip of plastic on both sides with metal foil. The plastic strip is 10 m long, 40 mm wide and 2.0×10^{-5} m thick, and is of relative permittivity 2.8. What is its capacitance, and what is the energy stored in it when the potential difference between the layers of metal foil is 500 V?

15-26 A mica capacitor consists of nine rectangular metal plates, measuring 30 mm by 20 mm, interleaved with sheets of mica of thickness 0.15 mm and relative permittivity 6.0. Calculate its capacitance.

15-27 Three parallel sheets of metal, each of area 0.15 m^2, are spaced apart by two large uniform sheets of insulator 2.0 mm thick and of relativity permittivity 3.0. What is the capacitance of the capacitor formed (a) by the centre as one electrode and the two outer sheets (joined together) as the other (b) taking the two outer sheets as the two electrodes, the centre sheet being left insulated?

15-28 Polythene has a relative permittivity of 3.0 and breaks down for electric field strengths in excess of 60 MV m^{-1}. If it is desired to construct a capacitor with polythene insulation of capacitance 1.0 μF and able to withstand p.d.s up to 1500 V, work out (a) the thickness of insulation required (b) the area of each plate (c) the energy per unit volume of polythene at the maximum design p.d. of 1500 V.

Draw a diagram to show the form of construction you would employ for this capacitor, and suggest values for the length and width of the sheet of polythene you would use.

15-29 Two square metal plates (0.10 m by 0.10 m) are arranged horizontally one above the other 2.5 mm apart. The lower plate is earthed, and the upper one is raised to a potential of 5.0 kV and then insulated. Calculate (a) the capacitance of the arrangement (b) the charge carried on each plate (c) the energy stored in the arrangement.

If now the upper plate is raised (keeping it insulated) until the gap between the plates is 5.0 mm, calculate now (d) the potential of the upper plate (e) the energy stored. By what means has the extra energy been supplied to the apparatus?

15-30 A capacitor consists of two parallel circular metal plates of diameter 250 mm placed 1.0 mm apart in air. These plates are connected to the cap and case of a leaf electrometer, and the system is charged by connecting it momentarily across a d.c. supply of 200 V. When the separation of the plates is increased to 4.0 mm, the deflection of the leaf increases until it indicates a p.d. of 700 V. Explain this observation, and calculate the capacitance of the leaf electrometer.

15-31 Two metal plates are arranged parallel to one another 1.50 mm apart; the capacitance of the capacitor so formed is 400 pF. The plates are permanently connected to the terminals of a storage battery of e.m.f. 24.0 V. Calculate (a) the charge on each plate (b) the energy stored in the capacitor.

The separation of the plates is now increased to 4.50 mm. Calculate (c) the new charge on each plate (d) the charge that flows through the battery, and state whether this charges or discharges the battery (e) the change in the energy stored in the *battery* (f) the change in the energy stored in the *capacitor* (g) the mechanical work done in increasing the plate separation (h) the average force with which one plate atracts the other.

15-32 A capacitor is filled with a poor quality insulator of resistivity 1.5×10^{12} Ω m and relative permittivity 4.0. This capacitor is charged by joining it to a 100 V supply and then isolated. Calculate (a) the time constant (RC) of the discharge process (b) the p.d. across the capacitor 60 s after being isolated from the supply.

15-33 An air capacitor of capacitance 300 pF is connected through a reed-switch alternately to a supply of 80 V and a microammeter. If the reed-switch performs 200 vibrations per second, calculate (a) the average current registered by the microammeter (b) the average rate at which energy is drawn from the supply (c) the average rate at which energy is delivered to the microammeter. Give an energy-flow diagram or chart to explain the energy conversions involved in this process.

15-34 Explain how you would use the apparatus described in the previous question to measure the relative permittivity of an insulating liquid.

15-35 A reed-switch capacitance meter is used to measure how the capacitance of a pair of parallel metal plates 0.25 m square varies with their separation d. The reed-switch connects the capacitor first to a supply of 25.0 V and then to a meter through which the capacitor is discharged. The frequency of the a.c. supply which drives the reed-switch is 300 Hz. The following values of the average current I through the meter are obtained for the values of d shown:

d/mm	1.5	3.0	4.5	6.0
I/μA	3.6	2.3	1.7	1.5

Plot a graph of I against $1/d$. Do you consider these results justify the assumption that the capacitance is inversely proportional to d? Explain why your graph does not pass through the origin. Measure the gradient of your graph, and use it to calculate a value for ϵ_0 the permittivity of air.

15-36 The effect of introducing a slab of insulator into the space between a pair of parallel plates is to reduce the field inside the insulator by a factor ϵ_r compared with its values in the air alongside the slab. Describe the physical processes by which this reduction of the field can be caused.

A pair of insulated parallel plates is set up 20 mm apart and a p.d. of 2.4 kV is applied momentarily between them. What is the electric field strength in the gap? If now a polythene slab 10 mm thick (for which ϵ_r is 2.4) is inserted midway between the plates, what is (a) the electric field strength in the air gaps (b) the electric field strength inside the polythene (c) the potential difference between the faces of the polythene slab (d) the potential difference across each air gap (e) the potential difference between the two plates?

15-37 The insulation of dry air breaks down in an electric field of strength 3.0 MV m^{-1}.
(a) What is the electric flux density in such a field?

(b) What is the surface density of charge on a metal object that would produce such a field?

15-38 A box made of insulating material contains three metal objects. These carry charges of $+ 3.5$ nC, $+ 2.5$ nC and $- 1.5$ nC. What is the total electric flux out through the surface of the box?

Fields near conductors

15-39 The following statements are all *false*. Explain the fallacy in each case, and describe how you would demonstrate it experimentally:
(a) 'There can be no electric field inside the case of a leaf electrometer with an earthed wire joined to its cap.'
(b) 'There can be no electric field inside a solid bar of conducting material.'
(c) 'There can be no electric field inside a closed metal box.'

15-40 Normally the air gap between a pair of parallel plates connected to a high-voltage supply is an almost perfect insulator, and a meter joined in series registers zero. Explain *two* ways in which you could introduce ions into the gap so that a current passes between them.

15-41 In order to discharge the surface of a plastic rod an experimenter moves it near a metal comb with very sharp teeth. Explain what is happening in this situation. Explain also why the method is not completely effective, being bound to leave some charge on the rod.

15-42 Some of the parts of a piece of electronic equipment are enclosed in aluminium cans. Explain why this is done.

15-43 An experimenter wishes to measure the charge carried by a small metallised sphere hanging on an insulating thread without letting the charge escape from the sphere. How do you suggest he should do this?

Radial fields

15-44 Two rain drops A and B falling side by side 10 mm apart carry charges of $+ 4.0$ pC and $- 5.0$ pC respectively, as in figure (a). What is the force with which one rain drop acts on the other? What is the electric field strength at a point X half way between them?

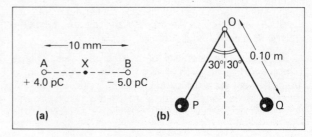

15-45 Two small conducting spheres P and Q each of mass 1.5×10^{-5} kg are suspended from the same point 0 by insulating threads 0.10 m long. When the spheres are charged (with equal charges) they come to rest with both threads inclined at 30° to the vertical, as shown in figure (b). Calculate the charges on the spheres.

If the charges on the two spheres were of the same sign but of *different* sizes, would the two threads now come to rest at *equal* angles to the vertical? Draw a diagram to show the kind of thing you would expect.

15-46 A uranium nucleus can be regarded as a spherical object of radius 2×10^{-14} m containing 92 protons (as well as neutrons). Estimate (a) the electric field strength (b) the electric potential at the surface of the nucleus.

If an α-particle (charge 2e) is close to the surface of the uranium nucleus, estimate (c) the electric force acting on it (d) the electrical potential energy it has in this position.

15-47 A small sphere bearing a charge of $+1.0$ nC is situated at a distance of 180 mm from another small sphere carrying a charge of $+4.0$ nC. Calculate the work that must be done in moving the first sphere to a point 60 mm from the second. What is the electrical force with which one sphere acts on the other in the latter position?

15-48 What is the capacitance of a metal sphere 6.0 m in diameter? If this sphere is charged to a potential of 5.0 MV by a van de Graaff machine, how much internal energy will be produced when it is earthed through a long resistance wire?

15-49 In a small van de Graaff generator the sphere at the top is of diameter 0.30 m; it is mounted on a plastic column whose resistance is 5.0×10^{10} Ω. If the current carried into the sphere by the belt is 3.0 μA, calculate (a) the final steady potential of the sphere (b) the final steady charge on the sphere (c) the electric field strength close to its surface.

15-50 In a small van de Graaff machine the sphere at the top is of diameter 0.20 m and reaches a potential of 3.0×10^5 V before discharging to an earthed object nearby. At what rate is charge being carried by the moving belt up into the sphere if there are 20 discharges per minute? If now a microammeter is permanently connected between the sphere and Earth, what current would you expect it to register?

15-51 An experimenter hangs a number of metallised spheres by nylon threads from a nylon washing line. He then charges each one in turn by connecting it for a moment by means of a long thin wire to the positive terminal of a 5.0 kV supply; the negative terminal of the supply is joined to a wire that goes into the Earth. The charge Q on each sphere is then measured with an electrometer, and the following results are obtained for the various values of radius r involved:

r/mm	10	14	22	28	40
Q/nC	5	8	12	15	22

Plot a graph to test the theoretical relationship for the capacitance C of an isolated sphere ($C = 4\pi\epsilon_0 r$). Use this graph to obtain a value for ϵ_0.

15-52 A length l of a single overhead supply cable carries a charge Q. Describe the pattern of lines of force that you would expect in the air near the cable. If the electric field strength is E at a distance r from the wire write down an expression for the total flux of the electric field out through a cylindrical surface of radius r concentric with the wire. According to Gauss's theorem this may be equated to the total charge contained inside the cylinder. From this equation work out an expression for the field E at the distance r from the wire.

Use your result to work out the value of E at a distance of 20 mm from a wire carrying a charge per unit length of 2.0 μC m^{-1}.

16 Magnetic forces

Data $g = 9.81$ N kg^{-1}
$e = 1.60 \times 10^{-19}$ C
$\mu_0 = 4\pi \times 10^{-7}$ N A^{-2}

Magnets

16-1 A bar magnet is supported on a block of wood floating in a tank of water. How will it behave if the Earth's field at the tank (a) is uniform (b) increases towards the northern end?

16-2 Two bar magnets are identical in size and appearance. Explain how, by observing their behaviour when suspended in the Earth's magnetic field, you would find (a) the N pole of each magnet (b) which magnet is the stronger.

16-3 A boy is tracing the magnetic field lines in the neighbourhood of a magnet. The three parts of the figure show features to be found in different parts of his line of force diagram.

In (a) he has two lines of force crossing; explain why this must be mistaken.

In (b) he has two adjacent lines of force running in opposite directions; what can you say about the field between them?

In (c) he has two lines of force that converge towards the top; what can you say about the field between these?

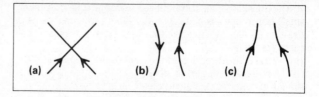

16-4 A bar magnet is laid on a table top in the magnetic meridian with its S pole pointing north. Draw a graph showing how you would expect the horizontal component of the magnetic field to vary along the axis of the magnet, starting from a point far to the north of the magnet and end at a point far to the south.

16-5 A straight wire carrying a steady current is laid on a horizontal table top. Describe how you would use a compass to find the direction of the current in the wire if the wire is laid (a) north and south (b) east and west.

The force on a current

16-6 A horizontal conductor of length 50 mm carrying a current of 3.0 A lies at right-angles to a horizontal magnetic field of flux density 0.50 T. What is the magnitude and line of action of the force acting on the conductor? Give a diagram showing how the directions of current, field and force might be related.

16-7 Currents of 3.0 A flow in turn in each of the conductors OA, OB, OC and OD, shown in the figure. The conductors are in a B-field of 0.15 T parallel to the plane of the diagram. What is the direction and magnitude of the force acting on each conductor.

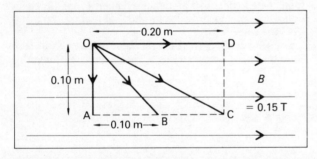

16-8 A straight wire 2.0 m in length lies at right-angles to a uniform magnetic field of 0.80 T. If the current in the wire is 60 A, calculate (a) the magnetic force acting on the wire (b) the mechanical power required to keep it moving at a speed of 10 m s^{-1} in a plane at right-angles to the field.

16-9 One end of a simple rectangular wire-loop current balance is inserted into a solenoid. A force of 3.0×10^{-3} N is found to act on this end when a current of 2.0 A is flowing in it. If the length of conductor forming the end of the wire-loop is 0.10 m, what is the magnetic flux density in the solenoid?

16-10 The magnetised needle of a small compass is pivoted so that it can swing freely in a horizontal plane. When a small bar magnet is placed on the table top with its axis on the line EOW, as shown in the figure, the compass needle turns through 30°.
(a) What is the direction of the field of the magnet at O?
(b) What is the direction of the horizontal component of the Earth's magnetic field at O?

(c) What is the direction of the resultant field at O? Draw a vector diagram showing the relation between these fields in magnitude and direction. Calculate the flux density of the field produced by the magnet at O, if the horizontal component of the Earth's magnetic field is of flux density 18 μT.

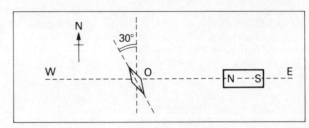

16-11 Referring to the previous question, a second small bar magnet is now placed with its axis on the same line EOW and is moved along it until the compass needle once more points north.
(a) What is the combined field of the two magnets at O?
(b) If one of the magnets is now turned over end for end, through what angle will the compass needle turn?

16-12 At a certain point the Earth's magnetic field is inclined at 70° to the horizontal and is of flux density 52 μT. Calculate for this point (a) the horizontal component of the field (b) the vertical component.

16-13 A plane rectangular coil of 20 turns, measuring 0.20 m by 0.10 m, carries a current of 0.50 A? What is the torque acting on this coil when it is placed with its plane parallel to a uniform magnetic field of flux density 40 mT? What is the torque if the coil is turned from its first position through an angle of (a) 60° (b) 90°?

16-14 What is the electromagnetic moment of the current-carrying coil referred to in the previous question?

16-15 A moving-coil meter has a coil of 100 turns wound on a square former ABCD of side 15 mm; it is pivoted about an axis through its centre parallel to AB and CD. It moves in a radial magnetic field of flux density 0.40 T. Calculate the force acting on each of the sides AB and CD when a current of 500 μA flows in the coil; show the direction of these forces on a diagram. What is the combined moment of these forces (a) about the central axis (b) about CD?

16-16 If a current of 500 μA gives full-scale deflection of 130° in the meter referred to in the previous question, calculate the suspension constant (i.e. the torque per unit twist) of the pair of hair springs.

16-17 A rectangular coil of 50 turns, measuring 20 mm by 15 mm, is suspended by two taut vertical metal strips so that it may turn about an axis through its centre parallel to one of its sides; it is placed in a uniform horizontal magnetic field. With no current in the coil it comes to rest with its plane parallel to the field. When there is a current of 10 mA in the coil it comes to rest with its plane at an angle of 40° to the direction of the field. If the stiffness (i.e. the torque per unit twist) of the pair of metal strips is 1.2×10^{-4} N m rad^{-1}, calculate the flux density of the field.

16-18 Two moving-coil meters X and Y are identical in all respects except the numbers of turns in their coils. X has a coil of 20 turns of resistance 12 Ω; Y has a coil of 80 turns of resistance 240 Ω. What is the ratio of the deflections of these meters when they are joined in a circuit (a) in series with one another (b) in parallel with one another?

16-19 Discuss what factors in the design of a moving-coil meter cause it to have high sensitivity.

16-20 A moving-coil meter of resistance 170 Ω is shunted by a resistance of 5.00 Ω. This is then joined in a circuit in series with a resistance of 1000 Ω and a lead-acid cell of e.m.f. 2.00 V, and the meter is found to give a deflection of 30.0 divisions. What is the calibration constant (i.e. the current per division) of the meter?

The force on a moving charge

16-21 What is the magnetic force acting on an electron moving with a velocity of 2.0×10^7 m s^{-1} at right-angles to a uniform magnetic field of 15 mT? Show on a diagram the direction of the force in relation to the directions of velocity and field.

16-22 A stream of helium nuclei each carrying a charge of 3.2×10^{-19} C is travelling with a velocity of 1.5×10^7 m s^{-1} at right angles to a uniform magnetic field of 2.0 T. What is the force on each particle? What effect does this force have on the path followed by the particle?

If the mass of a helium nucleus is 6.6×10^{-27} kg, what is the acceleration of a particle?

16-23 What is the magnetic force that acts on an electron of speed 8.0×10^6 m s^{-1} in a uniform magnetic field of flux density 1.5 mT as shown in the figure, if it is travelling (a) in the direction OA (b) in the direction OB (c) in the direction OC. What is the direction of the force in each case?

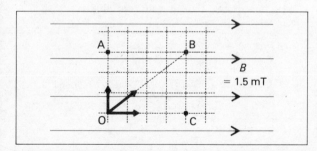

16-24 A cathode-ray tube is placed inside a solenoid so that the whole tube is in a uniform field of 10 mT parallel to the axis of the tube. What force acts on an electron of speed 1.2 $\times 10^7$ m s^{-1} travelling (a) along the axis of the tube (b) at an angle of 15° to the axis? Show on a diagram the direction of the force in relation to the directions of field and velocity.

In case (b) describe what the subsequent path of the electron will be.

16-25 When a Hall probe is placed with the germanium slice at an angle of 90° to a uniform magnetic field, a Hall p.d. of 0.20 V is measured. What p.d. would be measured if the probe is turned so that the angle is (a) 50° (b) 0°?

16-26 An electric current of 5.0 A is passing through a copper wire of diameter 1.0 mm. Calculate (a) the drift speed of the free electrons in it if their density in the copper is 1.0×10^{29} m^{-3} (b) the magnetic force acting on each electron if the wire is at right-angles to a magnetic field of 1.5 T (c) the strength of the electric field that would produce a force equal and opposite to the magnetic force on each electron (d) the Hall p.d. that could therefore arise between diametrically opposite points on the surface of the wire.

16-27 How many free electrons are there in a 0.40 m length of the wire referred to in the previous question? Hence use the result of part (b) of that question to calculate the total magnetic force acting on all the free electrons in the length of wire. Compare this with the force F given by the expression $F = BIl$ for this length of wire.

B-fields of coils and wires

16-28 If the flux density of the magnetic field at the centre of a solenoid is 20 mT, what can you say about the flux density at the ends of the solenoid? How would you test your prediction experimentally?

16-29 Calculate the flux density of the magnetic field at the centre of a solenoid of length 0.30 m uniformly wound with 380 turns of wire when the current in it is 1.5 A.

16-30 A long single-layer solenoid is wound with insulated wire of diameter 1.6 mm, the turns being laid down closely touching one another. What current is required to produce at its centre a field of flux density 5.5×10^{-3} T?

16-31 What is the flux density of the magnetic field produced at the centre of a flat circular coil of 20 turns of diameter 0.25 m by a current of 0.50 A?

If this coil is set with its axis horizontal in the magnetic meridian, what are the maximum and minimum values of the horizontal flux density at its centre produced by this current if the Earth's horizontal magnetic field is 18 µT?

16-32 A long solenoid is set so that its axis is horizontal and at right-angles to the magnetic meridian; the solenoid has 800 turns and is of length 0.40 m. If the horizontal component of the Earth's magnetic field is 18 µT, what is the direction of the resultant field at the centre of the solenoid when a current of 6.0 mA is switched on in it?

16-33 In a coaxial cable the return current in the outer cylindrical sheath produces no magnetic field inside it; the field between the two conductors is caused entirely by the current in the inner conductor. Such a cable consists of an inner wire 1.0 mm in diameter and an outer sheath 7.0 mm in diameter. If this cable carries a current of 5.0 A, tabulate values of the flux density B for various values of the radius r from the surface of the inner wire to the outer sheath; and plot a graph of B against r. Measure the gradient of this graph for $r = 2.0$ mm.

16-34 A long straight wire is arranged vertically so that it passes through a slot in a horizontal table top, and a current is driven through it in a downward direction. The horizontal component of the Earth's magnetic field on the table top is known to be 18 µT. If a neutral point (i.e. a point where the horizontal component of the magnetic field is zero) is found 100 mm from the wire, calculate the current in the wire. In which direction from the wire is the neutral point situated?

In what direction would a small compass point when it is placed on the table top 100 mm magnetic north of the wire?

16-35 A long straight wire carries a current of 6.0 A.
(a) What is the flux density produced by this current at a distance of 50 mm in air from the wire? Indicate the direction of the flux density on a diagram.
(b) What is the force acting on a length of 1.0 m of a similar wire carrying a current of 2.5 A, if this is 50 mm from the first and parallel to it? What is its direction?
(c) Deduce a general expression for the force per unit length with which one long straight wire acts on another such wire parallel to it. Explain how this result is used in the definition of the *ampere*.

16-36 A twin cable supplying electric power to a factory building consists of a pair of parallel straight wires 120 mm apart. If the current in the cable at one instant is 65 A, what is the force which one wire exerts on a length of 1.0 m of the other wire? Show its direction on a diagram.

16-37 In the circuit shown X, Y and Z are three *identical* components joined to a d.c. supply.
(a) If X, Y and Z are solenoids, explain (i) how the magnetic fields inside them are related (ii) how the rates of production of internal energy in them are related.
(b) If X, Y and Z are capacitors, explain (i) how the electric fields in them are related (ii) how the electrical energies stored in them are related.

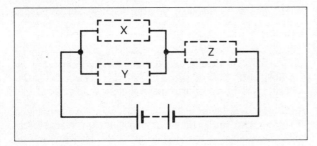

Induced e.m.f.s

16-38 Consider a single conductor forming part of one side of a coil in the armature of a d.c. dynamo. As the armature rotates this conductor moves across the magnetic field produced by the field magnet of the dynamo. Draw this conductor and indicate the direction of the magnetic field (passing perpendicularly through the diagram) and the direction of motion of the conductor through the field. Then insert on the diagram, in the right directions, (a) the force F_p

that acts on a proton in the conductor (b) the force F_e that acts on an electron in the conductor (c) the positive and negative ends of the conductor (d) the current I in the conductor when the armature is connected in a complete circuit (e) the magnetic force F_m that acts on the conductor because of the current in it (f) the force F_o with which the armature slot pushes the conductor to keep it moving at a steady speed.

Explain how the conversion of mechanical energy to electrical energy takes place in a dynamo.

16-39 The wing span of a Concorde aircraft is 25.6 m.
(a) What is the area swept out per second by this wing span when the aircraft is flying horizontally at 660 m s^{-1}?
(b) What is the potential difference between its wing tips if the vertical component of the Earth's magnetic field is 50 µT? Which wing tip is positive in the northern hemisphere?
(c) If a wire is connected internally between the wingtips, would any current flow in it? Explain your reasoning.

16-40 A railway train is travelling due north at 160 km per hour. Calculate the e.m.f. induced in an axle of length 2.0 m at a place where the Earth's magnetic field is 50 µT inclined at 65° to the horizontal.

What would you need to do to draw power from this e.m.f. for use inside the train (e.g. to heat a compartment)? Would the e.m.f. and the power be affected by the direction in which the train is travelling?

16-41 A wheel with metal spokes as shown in the figure is set spinning clockwise about its axle O at 4.0 revolutions per second; there is a uniform magnetic field of flux density 20 mT at right-angles to the plane of the wheel as shown.
(a) If the axle is earthed, calculate the potential of (i) a point A at a distance of 0.20 m from the centre (ii) a point B 0.10 m from the centre.
(b) What is the potential difference (i) between A and B (ii) between A and a point A' symmetrically placed diametrically opposite A (iii) between B and a point B' which is also 0.10 m from the centre?
(c) If the rim of the wheel is made of metal, would any current flow in this? Explain your reasoning.

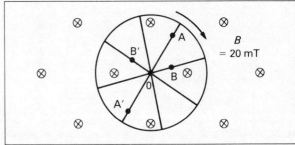

16-42 An electric fan has four blades each 150 mm long; it turns about a horizontal axle that points magnetic north. If the horizontal component of the Earth's field is 18 µT, and the speed of rotation of the fan is 2400 revolutions per minute, calculate the e.m.f. induced (a) between the axle and the tip of any one blade (b) between the tips of two blades directly opposite one another (c) between the tips of two adjacent blades.

16-43 A copper disc of radius 50 mm is turned about its central axis at 80 revolutions per second; the disc is at right-angles to a magnetic field of flux density 1.2 T. Sliding contacts bear on the centre and rim of the disc and through these a circuit is connected up whose total resistance (including that of the disc) is 0.40 Ω. Calculate (a) the current in the circuit (b) the rate of conversion of mechanical energy to electrical energy in the disc.

16-44 A rectangular coil of 60 turns measuring 200 mm by 300 mm is rotated about an axis in its own plane at right-angles to a magnetic field of flux density 0.15 T. If the speed of rotation is 300 revolutions per minute, calculate the peak value of the induced e.m.f. Draw a diagram to make clear the position of the coil in which this peak value is produced.

If in this position the current in the coil is 1.5 A, what is the torque needed to maintain this speed of rotation?

Energy conversions

16-45 A small d.c. dynamo is connected to a light bulb. A thread is wrapped round the shaft of the dynamo and a weight is hung on this so that the dynamo is rotated by the weight falling to the floor. Describe the energy transformations involved, and explain the physical process by which each transformation of energy is brought about. What exactly is the *electrical energy* that appears in this circuit?

16-46 An electric motor takes a power of 200 W from the mains when it is running at a steady speed of 3000 revolutions per minute with no load connected. Calculate the frictional torque exerted by its bearings and air resistance.

16-47 An electric motor (with field coils in parallel with armature) is run on a 24 V d.c. supply. When it is producing 120 W of mechanical power the armature current is 5.50 A. Calculate (a) the e.m.f. induced in the armature (b) the resistance of the armature.

16-48 When the circuit shown in the figure is connected up to a d.c. supply, the reading of the ammeter (whose resistance is negligible) is 4.0 A. The resistances of the other three components are as shown; the motor is of the kind that

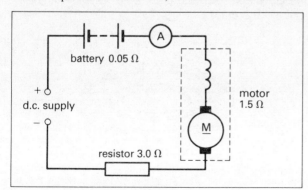

has its field coil in series with its armature, and their combined resistance is 1.5 Ω. When a high-resistance voltmeter is joined across the battery it reads 12.8 V; joined across the motor (armature *and* field coil) it reads 24 V. Calculate (a) the rate at which electrical energy is converted in each of the components (b) the rate at which internal energy is produced in each component. (c) How do you account for the differences (if any) between (a) and (b) in each case.

16-49 One of the coils in the armature of a motor has 25 turns each of area 5.0×10^{-3} m^2; the coil is of resistance 0.50 Ω and carries a current of 2.0 A.
(a) Draw a diagram showing the position of the coil in relation to the magnetic field when the torque acting on it is a maximum.
(b) If this maximum torque is 0.20 N m, what is the flux density of the magnetic field in which the coil moves?
(c) What is the e.m.f. induced in the coil in this position if its speed of rotation is 3000 revolutions per minute?
(d) What potential difference must be applied across the coil in this position to keep it turning at a steady speed?

16-50 A d.c. motor is designed with its field coil (of resistance 16 Ω) joined in parallel with its armature (of resistance 0.60 Ω). Connected to a supply of 24 V it takes, when running at a steady speed, a current of 6.5 A. Calculate (a) the current in the field coils (b) the current in the armature (c) the e.m.f. induced in the armature, (d) the mechanical power produced in the armature (e) the electrical power supplied to the whole motor (f) the efficiency of the motor.

16-51 The magnetic field in the electric motor of a toy train is provided by a permanent magnet. The motor is designed to work from a 12 V d.c. supply, and its armature has a resistance of 5.0 Ω. Running light on level track it reaches a speed of 0.45 m s^{-1} and takes a current of 0.60 A. When a carriage is hitched up to the engine, the maximum speed it reaches is 0.30 m s^{-1}. What current does it now take?

What is the maximum current this motor would take when starting from rest?

16-52 An electric motor has its field coils in parallel with its armature; it is running light at a steady speed on a d.c. supply. Explain fully what happens in the following circumstances.
(a) The load driven by the motor increases.
(b) The connections of the d.c. supply are reversed.

When a resistance is inserted in a series with the field coil only, the speed of the motor increases. Explain why this happens.

16-53 An experimenter uses a small electric motor on a 12 V d.c. supply to drive a load; the load is joined to the motor by means of a belt running over a pulley wheel on the motor shaft. When he switches on the motor, the 10 A fuse in the circuit melts. But if he allows the motor to get up to near maximum speed with the belt off, and then slides the belt on, the fuse does not melt and the motor is able to drive the load. Explain why this is so.

Explain how you would use a variable resistor to enable this motor to be started with the belt in place. What value of resistance would you choose?

16-54 In figure (a) the magnet is being pushed, N pole first, towards the square copper plate. In figure (b) the magnet is being moved sideways above the horizontal copper plate with its N pole a constant distance above it.

In each case describe with the aid of diagrams the pattern of currents you would expect to find in the copper plate, and show the directions of the magnetic fields caused by these currents. Explain in each case how the magnetic forces acting (a) on the currents (b) on the magnet tend to prevent the relative motion of the two.

In both cases internal energy is produced inside the copper plates. Explain where this energy comes from.

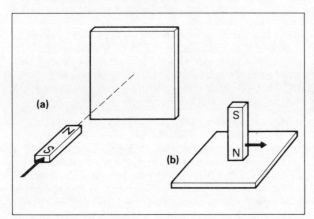

16-55 A bar magnet is suspended in a cradle on the end of a thread of cotton so that it can oscillate in the Earth's magnetic field. Describe and explain what happens when (a) a sheet of glass (b) a sheet of copper is placed horizontally a short distance below the oscillating magnet.

16-56 When the coil of a moving-coil light-spot meter is swinging freely it may be brought to rest by short-circuiting the coil. Explain why this happens.

The coil in the usual pivoted type of moving-coil meter does not oscillate at all when it is disturbed; but in the taut-suspension type with nothing connected to its coil it oscillates freely when disturbed. What difference in the construction of the two types of instrument causes this difference of behaviour?

16-57 A boy has a train set designed to run on a 12 V d.c. supply. However, all he has available is a 24 V d.c. supply. He reasons that all he has to do to adapt his set to the new supply is to double the resistance of the engines; this he does by joining in series with each engine armature a resistance equal to that of the armature. He then finds that, while the engines start with normal acceleration, they reach much greater maximum speeds. How do you account for this behaviour (including the fact that the engines start normally)?

17 Changing magnetic fields

Data $\mu_0 = 4\pi \times 10^{-7}$ H m^{-1} (or N A^{-2})

More induced e.m.f.s

17-1 Two solenoids are mounted coaxially one inside the other. One of them is connected to a sensitive light-spot meter, and the other through an ammeter to a variable d.c. supply. When the current in the latter solenoid is increased from zero to 0.50 A at a uniform rate in 10 s, the light-spot is deflected 30 mm to the right. What deflection would you expect (a) if the current is increased to 0.50 A in 20 s (b) if the current is decreased from 0.50 A to zero in 5.0 s (c) if, with the current steady at 0.50 A, the inner solenoid is moved slowly right out of the other one in 10 s at such a speed as to keep the deflection steady.

17-2 Two coils of insulated wire are wound one on top of the other on a wooden rod. The outer coil is connected to a sinusoidal a.c. supply and the inner one to a resistor. Draw diagrams (looking down the length of the rod) showing the directions of the currents in the two coils when the current in the outer one is (a) clockwise and increasing (b) clockwise and descreasing (c) anticlockwise and increasing. Sketch a pair of graphs showing how the currents in the two coils vary with time.

If the supply frequency is varied while keeping the peak value of the alternating current the same, how will this affect the alternating current in the resistor?

17-3 A circular coil and a copper ring are placed as in the figure flat on a table top. The current in the coil is increased from zero to 4.0 A in 2.0 s at a steady rate. Show on a diagram the directions of the currents in the coil and in the ring. How will (a) the current in the ring (b) the total energy conve:ted in the ring be affected if the current in the coil is increased from zero (i) to 2.0 A in 2.0 s (ii) to 2.0 A in 1.0 s (iii) to 4.0 A in 1.0 s?

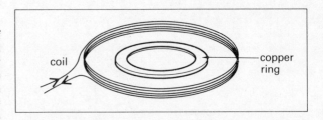

17-4 An experimenter makes a series of small search coils of varying numbers of turns and of cross-sectional shapes and sizes as shown in the figure. These are mounted in turn in a high-frequency alternating magnetic field in a solenoid, the planes of the coils all being perpendicular to the field. With coil (a) the peak value of the induced e.m.f. is 0.20 V. What will it be with the other coils?

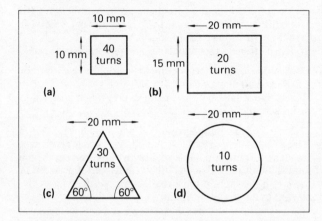

17-5 A circular copper ring of diameter 0.20 m and resistance 0.010 Ω is placed with its plane perpendicular to a magnetic field. If the field is changing at a rate of 2.0 mT s^{-1}, calculate (a) the e.m.f. (b) the current in the ring.

17-6 A closed circular loop of wire of diameter 120 mm is mounted inside a solenoid with its plane at right-angles to the axis of the solenoid. The current in the solenoid is varied so that the magnetic field inside it is changing at a rate of 2.0 × 10^{-2} T s^{-1}. Calculate (a) the e.m.f. induced in the loop (b) the rate of production of internal energy in the loop if its resistance is 0.50 Ω· (c) the potential difference between opposite ends of a diameter of the loop.

Draw two diagrams showing the loop in the plane of the paper and the magnetic field at right-angles to it, and indicate on these the directions of the current in the loop and the forces on it (i) if the magnetic field is increasing (ii) if the field is decreasing.

17-7 A search coil placed with its axis parallel to a magnetic field of peak value 1.2 mT oscillating sinusoidally at a frequency of 500 Hz has an alternating e.m.f. induced in it of peak value 2.0 V. In another similar field of frequency 20 kHz the peak value of the induced e.m.f. is 3.0 V. What is the peak value of the latter field?

17-8 An alternating current of 3.0 A at a frequency of 50 Hz is passed through 50 turns of a helical spring stretched to a length of 0.80 m. A search coil with its axis parallel to that of the spring is mounted inside the spring near its centre; it is connected to an oscilloscope, and with suitable settings of the controls the vertical height of the trace depicting the induced e.m.f. in the search coil is 40 mm. What height of trace would be obtained in the following alternative circumstances:

(a) The frequency of the current in the spring is 100 Hz.
(b) The 50 turns of the spring are stretched to 1.60 m.
(c) Only 25 turns are used, stretched to 0.80 m.
(d) The axis of the search coil makes an angle of 60° with that of the spring.
(e) The search coil is moved to the end of the spring, coplanar with the end turn, but still coaxial.

17-9 A search coil of 5000 turns of average area 1.00 × 10^{-4} m^2 is mounted in a magnetic field of peak value 3.5 mT oscillating sinusoidally at a frequency of 200 Hz. Explain at what moments the induced e.m.f. in the search coil reaches its peak value. Calculate this peak value (a) when the axis of the search coil is parallel to the field (b) when it is inclined at an angle of 70° to the field.

17-10 Two overhead electric supply cables running to a house are 0.20 m apart. What is the magnetic field mid-way between them when the current supplied is 20 A? A search coil of 5000 turns of mean area 1.0 × 10^{-4} m^2 is mounted in this position coplanar with the pair of cables. What is the peak value of the e.m.f. induced in the coil when the peak value of the 50 Hz current in the cables is 20 A?

Magnetic flux

17-11 The average flux density of the Earth's magnetic field over the British Isles is 53 μT inclined downwards at an angle of 70° to the horizontal. Calculate the total flux of the field through the British Isles if these are of area 3.0 × 10^{11} m^2.

17-12 A large electromagnet has circular pole pieces of diameter 0.20 m. The total flux produced by the magnet is 0.050 Wb. Calculate the average flux density of the field between the pole pieces.

17-13 A solenoid of length 0.30 m is uniformly wound with 400 turns of wire. If a current of 1.5 A is flowing in it, what is the flux density of the magnetic field at its centre? If the solenoid is of square cross-section (20 mm by 20 mm), what is the total magnetic flux through its centre?

17-14 A bar magnet of circular cross-section 12 mm in diameter is equipped with a close-fitting coil of 50 turns. This is placed at the centre of the magnet and connected to a sensitive meter. The magnet is then removed from the coil in such manner as to keep the reading of the meter steady at 200 μV; this is found to take 4.0 s. Calculate the flux density at the centre of the magnet.

Inductance

17-15 A current of 5.0 A in a coil of 25 turns is found to produce a magnetic flux Φ in its core of 3.0 × 10^{-5} Wb. Calculate the inductance of the coil.

17-16 The current I in a certain coil grows intially at a rate of 20 A s^{-1} when a steady p.d. of 5.0 V is applied to it. What is its inductance?

17-17 When the current in a certain coil is 5.0 A, the total magnetic flux ($N\Phi$) linked with it is 0.40 Wb. Calculate (a) the inductance of the coil (b) the average p.d. that must be applied to the coil to reverse the current in it in 2.0 ms.

17-18 A coil of inductance 30 H is connected directly to a 240 V d.c. supply. The circuit, which is of negligible resistance contains a 13 A fuse. How long after completing the circuit does the fuse melt? Describe with the aid of graphs how the current in the coil and the p.d. across it vary from the moment when the connection is first made until *after* the fuse has melted.

17-19 The pair of line deflector coils attached to the neck of a television tube carry a current that follows a saw-toothed waveform as shown in the figure. From A to B the current changes at a uniform rate from -0.25 A to $+0.25$ A in 62 μs; from B to C the current returns to -0.25 A in 2.0 μs. If the pair of coils is of inductance 6.5 mH and of negligible resistance, calculate the p.d. across the coils at each stage of the waveform, and draw a sketch-graph of the waveform of p.d. across the coils.

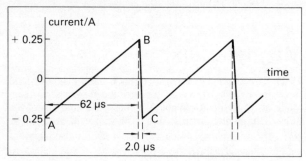

17-20 An air-cored coil of inductance 150 mH and resistance 75 Ω is connected directly across a 12 V lead-acid battery. Calculate (a) the initial rate of growth of current (b) the final steady current (c) the time constant of the coil (d) the current 20 μs after connecting the coil.

17-21 The flux $N\Phi$ linking a coil of N turns because of the current I in *another* coil nearby is given by $N\Phi = MI$, where M is a constant called the *mutual inductance* of the two coils. If M for a certain pair of coils is 0.40 H, what is the e.m.f. induced in the second coil when the current is growing at 25 A s^{-1} in the first coil?

17-22 A lamp marked '12 V, 0.5 A' is joined in series with a coil of many turns and connected to a 12 V supply. When a d.c. supply is used, the lamp shines at full brightness; but when an a.c. supply is used the brightness of the lamp is considerably less. Outline the physical principles that enable this observation to be explained.

When the a.c. supply is in use what would you expect to happen (a) if the frequency of the supply is increased while keeping the p.d. at 12 V (b) if a laminated iron core is inserted into the coil?

17-23 A sinusoidal alternating current of peak value 1.2 A and of frequency 50 Hz flows in a coil of inductance 0.25 H. Calculate (a) the maximum flux ($N\Phi$) linking the coil (b) the peak value of the e.m.f. induced in the coil.

17-24 An air-cored coil is joined in series with an a.c. meter to a signal generator that gives a sinusoidal output of variable frequency f at an r.m.s. potential difference V_{rms} of 1.00 V. The following readings of the r.m.s. current I_{rms} are obtained.

f/Hz	40	60	100	200	400	600	1000	1400
I_{rms}/mA	33	32	30	24	15	11	7	5

Plot a graph of Z ($= V_{rms}/I_{rms}$) against f. Use your graph to decide for what range of frequencies this coil behaves (to within 10%) effectively as (a) a 'pure' inductance ($Z \propto f$) (b) a 'pure' resistance (Z constant).

Hence estimate the inductance and resistance of the coil.

17-25 What is the energy required to establish a current of 1.5 A in a coil of inductance 1.6 H?

Describe what becomes of this energy when a switch in series with the coil is suddenly opened.

17-26 A coil of inductance 8.0 H and resistance 4.0 Ω is joined to a battery of e.m.f. 12 V and of negligible resistance. Calculate (a) the initial rate of growth of the current (b) the final value of the current (c) the average e.m.f. induced in the coil if a switch in the circuit is opened that causes the current to fall to zero in 10 ms.

An experimenter fears that this induced e.m.f. may cause breakdown of the insulation of the coil; so he joins a semiconductor diode (which conducts current in one direction only) across it. (d) Draw a circuit diagram showing which way the diode must be connected, and (e) calculate the energy transformed to internal energy in the diode and coil during switching off.

17-27 Estimate the inductance of a cylindrical air-cored solenoid of diameter 50 mm and length 0.40 m uniformly wound with 2500 turns of wire.

17-28 Certain conductors can be made superconducting by cooling them to the temperature of liquid helium; they then have zero resistance. However if the magnetic field they are in grows beyond a certain point the resistance of the conductor becomes normal and its reistance returns. A superconducting solenoid in liquid helium is connected to a cell of e.m.f. 2.0 V and of negligible resistance. After 20 s the current has grown to 4.0 A, and the solenoid acquires resistance. The energy stored in it is then converted to internal energy and a quantity of helium boils off. Calculate (a) the inductance of the solenoid (b) the maximum energy stored in it (c) the volume of helium at s.t.p. that is evolved, taking the s.l.h. of helium as 22 kJ kg^{-1} and the density of helium gas at s.t.p. as 0.18 kg m^{-3}.

The transformer

17-29 Describe with the aid of a diagram how you would put together a simple transformer to produce an a.c. supply of 50 V from a 12 V a.c. supply. You have available a pair of C-cores and a range of coils that fit them, and you are told to use a coil of 100 turns for the primary.

If the terminals of the secondary coil are left unconnected, what will be the effect of an increase in the supply frequency (at 50 V) (a) on the primary current (b) on the output p.d.?

17-30 A transformer has a primary coil of 2000 turns and a secondary of 50 turns. The primary is connected to an a.c. supply of 240 V, and the secondary to a resistance of 2.0 Ω. Estimate the current (a) in the secondary (b) in the primary. Explain why these answers are only estimates, and whether the actual currents in each case are likely to more or less than the values you give.

17-31 A coil of many turns is joined in series with a suitable meter to an a.c. supply. Explain the following observations.
(a) When a laminated iron core is inserted into the coil the current *decreases,* but the temperature of the iron core does not perceptibly increase.
(b) When a solid copper core is inserted into the coil the current *increases*, and the copper core becomes hot.

17-32 Two coils P and Q are mounted on a closed pair of C-cores. Q is left open-circuited, while P is connected to a d.c. supply that causes the current in it to increase at a uniform rate from 0 to 7.5 A in 3.0 s. During this period there is a potential difference of 0.050 V across P and 0.60 V across Q. Calculate (a) the inductance of P (b) the turns ratio of Q to P (c) the inductance of Q.

17-33 In a certain transformer used in a 50 Hz mains circuit the rate of production of internal energy under normal load conditions is 30 W in the iron core of the transformer and 20 W in the copper coils wound on it. What would you expect these quantities to become (a) if the current taken from the secondary increases by 10% (b) if the frequency and primary p.d. increase by 10%, but the currents remain the same.

17-34 A transformer draws a current of 0.25 A from the 240 V a.c. mains supply. Its output produces a current of 0.50 A at a p.d. of 100 V. The primary coil of the transformer is of resistance 70 Ω, and the secondary coil of resistance 10 Ω. Calculate the following quantities (assuming the currents are in phase with the p.d.s):
(a) the power input
(b) the power output
(c) the efficiency of the transformer
(d) the rate of production of internal energy in each coil
(e) the rate of production of internal energy in the iron core.

Magnetic materials

17-35 A closed rectangular loop of wire may be made superconducting by cooling it to the temperature of liquid helium. Explain what happens (with diagrams showing the relative directions) to the current in the loop, and the forces acting on it in the following circumstances.
(a) The loop is cooled to the temperature of liquid helium in zero magnetic field; and then the N pole of a long bar magnet is brought up towards it. Work is done in this process; what form of energy does it produce?
(b) The loop is placed in a large magnetic field at room temperature, and is then cooled to the temperature of liquid helium; finally the field is reduced to zero.

Explain how a superconducting loop provides a model for a molecule, in case (a) of a diamagnetic substance, and in case (b) of a paramagnetic substance.

17-36 Describe the behaviour of a typical ferromagnetic material in a large oscillating magnetic field. Represent this behaviour by means of a suitable free-hand graph. Explain with the aid of your graph the difference between the properties required in a material to make (a) a transformer core and (b) a permanent magnet.

17-37 Explain how the pattern of magnetic domains may be revealed on the surface of a crystal of iron under a microscope. Sketch a possible pattern of domains, and explain how this pattern changes as the magnetic field round the crystal is gradually increased.

17-38 An iron ring of mean circumference 0.40 m is uniformly wound with 2000 turns of wire. A current of 0.20 A in this coil is found to produce a flux density in the iron ring of 0.90 T. Calculate the relative permeability of the iron.

17-39 The iron core of a small mains transformer is of square cross-section (20 mm by 20 mm), and the other dimensions of the core are as shown in the figure. The relative permeability of the iron is 1500. Estimate the inductance (a) of the primary coil on its own (b) of the secondary coil on its own (c) of the two coils joined in series (in such a way that their magnetic fields *add*).

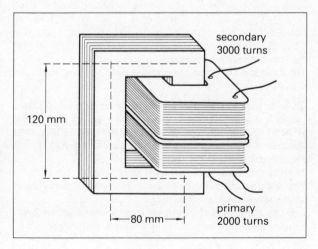

secondary
3000 turns

120 mm

80 mm

primary
2000 turns

18 Electrons and quanta

Data $c = 3.00 \times 10^8$ m s^{-1}
$e = 1.60 \times 10^{-19}$ C
mass of an electron $m_e = 9.11 \times 10^{-31}$ kg
hence, specific charge of an electron e/m_e
$\qquad = -1.76 \times 10^{11}$ C kg^{-1}
mass of a proton $m_p = 1.67 \times 10^{-27}$ kg
Planck constant $h = 6.63 \times 10^{-34}$ J s

The properties of electrons

18-1 Explain how experiments with a simple thermionic tube (e.g. a Maltese cross tube) lead us to believe
(a) that 'rays' other than light are emitted by a heated cathode
(b) that the rays travel from the cathode approximately in straight lines
(c) that the rays carry negative charge
(d) that the rays carry energy.

18-2 In a television picture tube 1.0×10^{16} electrons strike the screen per second. The accelerating potential difference used is 16 kV. Calculate (a) the current in the tube (b) the power converted within it.

18-3 An electron is accelerated from rest through a potential difference of 200 V. Find the speed it acquires.

18-4 It is found that there is a small current through a thermionic tube with an anode and a heated cathode even when the anode is slightly negative with respect to the cathode. But this current ceases when the anode potential reaches -0.60 V with respect to the cathode. Calculate (a) the maximum energy with which electrons are emitted by the cathode (b) their maximum speed as they leave the cathode.

18-5 A thermionic tube contains an anode and a heated cathode which is earthed. When the anode is at a potential of +50 V, the current through the tube is 25 mA. The distance between cathode and anode is 40 mm. Use this data to calculate the following:
(a) the speed of the electrons as they strike the anode
(b) the time taken by the electrons to cross the gap between cathode and anode if their acceleration is uniform
(c) the number of electrons striking the anode per second
(d) the number of electrons in the gap at any moment

18-6 The current I through a thermionic tube with a heated cathode (which is earthed) for varius values of the potential V of the anode is given in the following table:

V/V	0	10	20	40	60	80	100	120	140	160
I/mA	0.5	4	13	41	82	128	153	160	163	164

Plot a graph of I against V, and make estimates of (a) the p.d. that must be applied to the tube to give a current of 100 mA (b) the maximum electron current that can be collected from this cathode (c) the resistance of the thermionic tube when $V = 50$ V.

According to one theory the relationship between I and V should be of the form: $I = kV^n$, where k and n are constants. Plot a graph of $\ln(I/\text{mA})$ against $\ln(V/\text{V})$, and use it to determine (d) over what range of p.d.s. the theory applies in this case, and (e) the value of n for this range.

18-7 A horizontal beam of electrons of speed 2.0×10^7 m s^{-1} passes through an electric field of 4.0×10^5 V m^{-1} which acts vertically downwards. What magnetic field would you need to apply over the same region as the electric field to maintain the beam horizontal? Show on a diagram what the direction of the magnetic field should be.

18-8 An electron beam of speed 1.8×10^7 m s^{-1} is observed to travel in a magnetic field in a circular arc of radius 0.15 m. Calculate the flux density of the field, and indicate on a diagram its direction in relation to the circular path of the beam.

18-9 Describe how you would measure the specific charge (e/m_e) of the electron, explaining the steps of calculation as well as the experiments involved. Explain carefully which of your experimental results indicate that all electrons are identical.

18-10 A beam of cathode rays is found to move in an arc of radius 0.45 m in a magnetic field of flux density 300 μ T. An electric field is then superimposed on the magnetic field in such a way as to oppose the magnetic deflection of the beam. When this field is adjusted to 6.9 kV m^{-1} the beam travels in a straight line. Calculate values for the speed and specific charge of the cathode rays.

18-11 In a fine-beam tube the electron beam is accelerated through a potential difference V, and is then bent into a circular path of radius r by a uniform magnetic field of flux density B. Deduce an expression giving r in terms of B, V and the specific charge (e/m_e) of the electron.
What will happen to the beam if (a) the B-field is doubled (b) the accelerating p.d. V is doubled?

18-12 A beam of electrons is accelerated in a cathode-ray tube by a potential difference of 1200 V. It then enters the space between two parallel metal plates 20 mm apart between which there is a p.d. of 40 V; the beam is initially parallel to the two plates and mid-way between them. How do you describe the path of the beam in the electric field? The beam is thus deflected and strikes one of the plates at a point A.
(a) show that the speed of the electrons as they enter the electric field is 2.06×10^7 m s^{-1}.
(b) What is their acceleration in the electric field?
(c) How long does an electron spend in the electric field before striking the metal plate?
(d) How far is the point A from the edge of the plate?

The cathode-ray oscilloscope

18-13 Describe the construction and principles of operation of a cathode-ray tube of the kind used in an oscilloscope. Explain what becomes of the electrons after they have been brought to rest on the screen, and make an estimate of the time of flight of the electrons from cathode to screen for a tube 0.25 m long using an accelerating p.d of 500 V.

18-14 Describe with the aid of an energy flow diagram the energy changes involved in the functioning of a cathode-ray tube, and state what form of energy is possessed by an electron (a) just after it has emerged from the cathode with negligible speed (b) just before it strikes the screen. Also (c) explain into what forms the energy of the electron is converted in the screen.

18-15 A sinusoidal alternating p.d. of frequency 50 Hz is connected to the Y-deflection terminals of an oscilloscope. Sketch the sort of trace you would expect to obtain on the screen if the usual saw-toothed time-base is applied to the X-plates (a) of frequency 10 Hz (b) of frequency 100 Hz. Sketch also the sort of trace that would be obtained if the time-base is disconnected and the same 50 Hz alternating p.d. is also applied to the X-plates.

18-16 A clarinet is played in front of a microphone joined to an oscilloscope, and the trace shown in the figure is obtained. The time base control is set at 1 ms div^{-1}. Calculate (a) the period of one complete oscillation (b) the frequency of the note.

What is the cause of the faint nearly horizontal line with which the trace starts at the left?

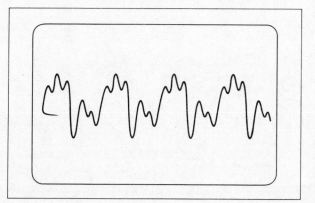

18-17 A small a.c. generator and a battery are joined in series, and the circuit is completed by a 100 Ω resistor. An oscilloscope connected across the resistor gives the trace

shown in the figure. (The horizontal line is the trace with no Y-input.)

The settings of the oscilloscope controls are: Y-input 'direct'; Y-amplifier control '1 V div $^{-1}$'; time base '10 ms div^{-1}'. If the generator and battery are of negligible internal resistance, calculate (a) the peak e.m.f. of the a.c. generator (b) the frequency of the a.c. generator (c) the e.m.f. of the battery (d) the maximum and minimum currents in the resistor.

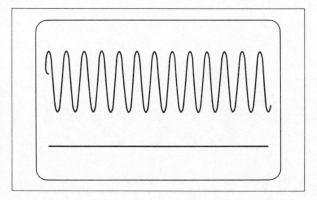

18-18 An oscilloscope is being used to investigate the p.d. between two points in the time base circuit of another instrument, and the trace shown below is obtained. The Y-amplifier control is set at 0.5 V div^{-1}, and the time base control at 0.1 ms div^{-1}. Estimate the following:
(a) the frequency of repetition of the waveform
(b) the rate of rise of p.d. during the straight part of the waveform
(c) the frequency of the damped oscillation that occurs during the 'fly-back' part of the trace
(d) the maximum change of p.d. during the 'fly-back'.

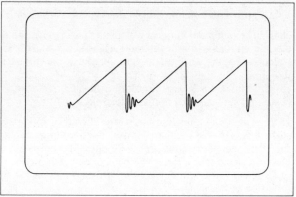

18-19 A capacitor of capacitance $0.22\ \mu\text{F}$ in a certain circuit has a p.d. of saw-toothed waveform applied to it as shown in the figure. From A to B the p.d. rises from -10 V to $+10$ V in 100 ms at a uniform rate, and from B to C it falls to -10 V again in 10 ms. Work out for each part of the waveform (a) the rate of change of p.d. across the capacitor (b) the current in the capacitor leads. Sketch the waveform of the current in the capacitor leads, and explain how you would connect up an oscilloscope to enable you to observe this waveform.

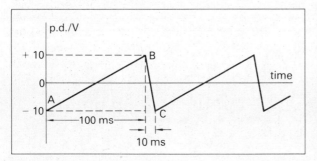

18-20 A beam of cathode rays of speed 1.5×10^7 m s^{-1} passes for 60 mm of its path through a uniform electric field of 6.5 kV m^{-1}, the direction of the beam being initially at right-angles to the field. Calculate
(a) the time spent by an electron in the field
(b) the acceleration of an electron in the field
(c) the resolved part of the velocity of the electrons in the direction of the field as they emerge from it. Then
(d) draw a vector diagram of velocities for the electrons as they emerge from the field, and
(e) work out the deflection of the beam by the field.

The photoelectric effect

18-21 The wavelengths of the two D lines in the spectrum of a sodium vapour lamp are both close to 5.9×10^{-7} m. What is the energy of one quantum of sodium D light?

A 200 W sodium vapour street light has an efficiency of 30% (i.e. 30% of the supplied energy is emitted as the D light). How many quanta of light does it emit per second?

18-22 The minimum frequency of light that will cause photoelectric emission from a lithium surface is 5.5×10^{14} Hz. Calculate (a) the work function of lithium. If the surface is illuminated by light of frequency 6.5×10^{14} Hz, calculate (b) the maximum energy (in eV) of the electrons emitted (c) the maximum speed of these electrons.

18-23 The work function of a freshly cleaned copper surface is 4.16 V. Calculate (a) the minimum frequency of the ultra-violet radiation that will cause the emission of electrons from the surface (b) the maximum energy (in eV) of the electrons emitted when the surface is illuminated by radiation of frequency 1.2×10^{15} Hz.

18-24 Explain under what circumstances electrons can be (a) emitted (b) collected from a cold metal surface illuminated by light.

An *uncharged* leaf electrometer has on its cap a clean zinc plate which is illuminated with ultra-violet radiation. Describe what you would expect to happen. To what extent would you expect the potential of the plate to change? Would this be sufficient to cause any deflection of the leaf? If not, what would you suggest doing to test your prediction?

18-25 Electrons are emitted from a certain metal plate for all wavelengths less than 6.0×10^{-7} m. What is the maximum energy (in eV) of the emitted electrons if the metal is illuminated with light of wavelength 5.0×10^{-7} m?

18-26 Ultra-violet radiation of wavelength 3.0×10^{-7} m illuminates a vacuum tube in which there are clean surfaces of zinc (work function 3.6 V) and tungsten (work function 4.5 V). Will electrons be emitted from either or both surfaces? If so, what is the maximum speed of the emitted electrons?

18-27 In an experiment with an illuminated photocell a small electron current passes through the tube even when the anode is made slightly negative with respect to the cathode. Explain this. The current falls to zero only when the reverse p.d. across the tube reaches a value V_s, which varies with the frequency f of the radiation used to illuminate the cathode. The graph in the figure shows the results obtained with two different photocells A and B.
(a) Write down the theoretical relation connecting V_s and f predicted by Einstein's photoelectric theory.
(b) Measure the gradient of the two graph lines.
(c) Hence calculate the Planck constant h.
(d) Measure the intercepts on the f axis for A and B.
(e) Hence calculate the work functions of the two cathodes.
(f) What is the significance of the fact that the two graph lines are parallel?

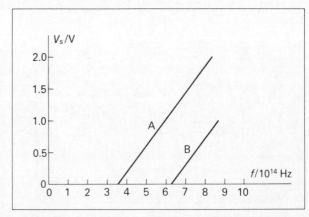

18-28 Show that for a particle of mass m travelling at low speeds (compared with the speed of light) the kinetic energy W and momentum p are connected by the relation
$$W = p^2/2m.$$
A particle with a single electronic charge is accelerated from rest in a vacuum tube through a potential difference of 10 V.

(a) Calculate its energy (i) in eV (ii) in J.
(b) Calculate the momentum of the particle (i) if it is a proton (ii) if it is an electron.
(c) Calculate the wavelength of the de Broglie wave of the particle (i) if it is a proton (ii) if it is an electron.

18-29 An experimenter wishes to investigate diffraction of electron waves by thin foils. He wants to use waves of wavelength 1.5×10^{-10} m.
(a) What momentum should the electrons have?
(b) What is the energy of these electrons?
(c) What p.d. should be used to accelerate the electrons?

18-30 An electron diffraction tube consists of a cathode-ray tube in which the electron beam passes through a very thin crystalline foil. The beams diffracted by the crystals form circles on the end face of the tube. The diameter d of one particular prominent circle in the diffraction pattern is measured for a range of values of the accelerating p.d. V, and the following values are obtained:

V/kV	1.5	2.0	3.0	4.5	6.0
d/mm	68	58	48	40	35

The theory of this experiment leads us to expect that $d \propto (1/\sqrt{V})$. Plot a graph of d against $1/\sqrt{V}$, and comment on the result. What p.d. would be required to give a diffraction circle of this kind of diameter 25 mm?

Energy levels

18-31 Explain what is meant by the collision of an electron with a gas atom being *elastic*. In a succession of such collisions what effect is there on the energy and speed of the electron?

What are the circumstances in which the collision of an electron with a gas atom may be inelastic? What energy conversions are involved in this case?

18-32 The energy W of the electron in a hydrogen atom can have only a number of sharply defined values. Taking the ionisation level of the electron as the zero of energy, the four lowest values of W are as follows:

quantum number n	1	2	3	4
energy W/eV	-13.6	-3.4	-1.5	-0.85

In this table $n = 1$ represents the ground state. According to Bohr's theory of the atom

$$\frac{1}{\sqrt{-W}} \propto n$$

Tabulate the values of $1/\sqrt{-W}$, and plot a graph of this against n to test this aspect of the theory. Use your graph to predict the values of W for $n = 5$ and $n = 6$.

What wavelength of radiation would be emitted by a transition of the electron between these last two levels, and in what part of the spectrum would you expect to find it?

18-33 The ionisation potential of hydrogen is 13.6 V. Describe how you would demonstrate this experimentally.
(a) Calculate the energy (in eV) and speed of the slowest electron that can ionise a hydrogen atom when it collides with it.
(b) Calculate the longest wavelength of electromagnetic radiation what could produce ionisation in hydrogen.

18-34 The lowest two excited states of a hydrogen atom are 10.2 eV and 12.1 eV above the ground state. Calculate *three* wavelengths of radiation that could be produced by transitions between these states and the ground state. In which parts of the spectrum would you expect to find these three wavelengths?

18-35 Three of the excitation potentials (above the ground state) of mercury vapour are 4.9 V (level X), 5.4 V (level Y), and 7.7 V (level Z). Transitions of the mercury atom are normally possible only (a) between the ground state and level X (b) between level X and level Z (c) between level Z and level Y. Calculate what wavelengths of radiation must be absorbed and emitted by an initially unexcited mercury atom in order to bring it to level Y. In each case state in what part of the spectrum the radiation occurs.

18-36 How do you account for the following experimental observations?
(a) Cool mercury vapour strongly absorbs the *ultraviolet* radiation from a mercury vapour lamp.
(b) Cool mercury vapour is completely transparent (non-absorbing) for the visible light from a mercury vapour lamp, provided the ultraviolet light is excluded by a glass filter.
(c) Cool mercury vapour partly absorbs the visible light from a mercury vapour lamp if the ultraviolet radiation from the lamp is also present.

Make an estimate of the lowest excitation potential of mercury vapour if the ultraviolet radiation from a mercury vapour lamp is of wavelength 254 nm.

Ion beams

18-37 In a mass spectrometer it is desired to focus singly charged ions of neon-20 with a magnetic field that causes them to travel in an arc of a circle of radius 0.40 m. If the ions are all travelling at the same speed, what would be the radii of the arcs followed in this magnetic field by (a) doubly charged ions of neon-20 (b) singly charged ions of neon-22?

18-38 In the previous question, if the ions of neon-20 are of mass 3.3×10^{-26} kg and are travelling at a speed of 2.0×10^5 m s^{-1}, what magnetic field is required?

18-39 A uniform electric field is maintained along the inside of a highly evacuated tube 5.0 m in length, the difference of potential between the ends being 2.0 MV. A proton is released from rest at the positive end of the tube. Calculate (a) the acceleration of the proton (b) the time taken for it to travel the length of the tube.

18-40 A beam of protons of speed 4.0×10^6 m s^{-1} moves in a circular path under the influence of a magnetic field of flux density 0.50 T. Calculate (a) the radius of the circular path (b) the time taken by a proton in completing one circular orbit (c) the frequency of revolution of the protons.

18-41 By repeating the calculation in the previous question using algebraic symbols rather than figures for the various quantities, show that the frequency of revolution of a proton in a given magnetic field is the same whatever its speed may be.

X-rays

18-42 Describe with the aid of a diagram an apparatus for producing X-rays. What effect would there be on the intensity and the penetrating power of the X-rays from an X-ray tube of an increase in (a) the potential difference across the tube (b) the current used to heat the cathode?

18-43 In a modern X-ray tube the anode consists of a plug of tungsten embedded in a substantial copper bar. Explain the reasons for the choice of these two materials. In a certain X-ray tube the accelerating p.d. is 50 kV and the beam current is 3.0 mA. Calculate the number of electrons striking the anode per second and the energy they impart to it per second.

18-44 In an industrial X-ray tube the target (anode) is kept at zero potential and is cooled by a continuous flow of water. The cathode is at a potential of -200 kV and its temperature is adjusted so that the current carried by the electron beam is 20 mA. What mass of water must pass through the cooling system per minute if its rise in temperature is not to exceed 40 K? (The s.h.c. of water is 4.2 kJ K^{-1} kg^{-1}.)

18-45 At what points in a typical living room containing the usual electrical equipment (lights, fire, television, etc) would you expect to find electromagnetic radiation of the following wavelengths: (a) 5×10^{-7} m (b) 3×10^{-10} m (c) 2×10^{-6} m (d) 1.5×10^3 m?

What are the photon energies (in eV) of each of these radiations?

18-46 What is the maximum *frequency* of the X-rays produced at the screen of a cathode-ray tube whose accelerating p.d. is 250 V? What is the range of *wavelengths* of X-rays emitted?

18-47 What is the minimum wavelength of the X-rays produced in an X-ray tube operated by a potential difference of 40 kV? Sketch a graph to show the general nature of the intensity of the X-ray beam produced for different wavelengths.

18-48 Calculate the p.d. across an X-ray tube in which the minimum wavelength of X-rays emitted is equal to the de Broglie wavelength of the electrons striking the target.

For the p.d.s normally used in X-ray tubes which of the two kinds of wave has the greater wavelength?

19 Probing the nucleus

Data $\epsilon_0 = 10^{-12}$ F m^{-1}
$c = 3.00 \times 10^8$ m s^{-1}
$e = 1.60 \times 10^{-19}$ C
mass of the electron $m_e = 9.11 \times 10^{-31}$ kg
hence, specific charge of the electron e/m_e
$\qquad = -1.76 \times 10^{11}$ C kg^{-1}
Planck constant $h = 6.63 \times 10^{-34}$ J s
unified atomic mass constant $\mathbf{m}_u = 1.66 \times 10^{-27}$ kg
Avogadro constant $L = 6.02 \times 10^{23}$ mol^{-1}
1 curie (Ci) $= 3.7 \times 10^{10}$ s^{-1}
1 year $= 3.16 \times 10^7$ s

Radioactivity

19-1 A radium source is mounted in total darkness above a rectangular photographic plate as shown in the figure; this is seen from the side in (a). After some time the plate is removed and developed, and is found to be fogged as shown in (b). Explain what kinds of radiation are responsible for the different areas of fogging, and how their properties give rise to the pattern shown.

To investigate these properties further an experimenter repeats this test with a photographic plate covered partly with a strip of thin card and partly with a strip of aluminium about 2.5 mm thick, as shown in (c). Explain what you would expect to find when this plate is developed. Would you expect to find any parts of it totally free from fogging?

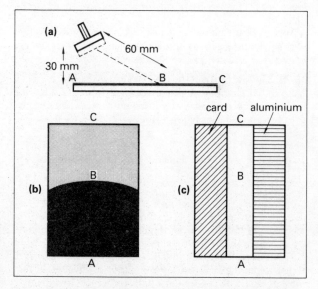

19-2 Describe how you would identify which types of nuclear radiation are emitted by an unknown radioactive source.

19-3 If an α-particle loses on average 30 eV of energy for each ion pair that it creates in a collision, how many ion pairs would you expect a 6.5 MeV α-particle to create? An α-particle source that emits particles of this energy is placed inside an ionisation chamber in such a way that all the ions created are collected by the electrodes in the chamber. If the source emits 2.5×10^5 particles per second, estimate the current through the ionisation chamber. Explain what assumption about the ions you are making.

Particle detectors

19-4 Describe some form of particle detector that will respond only to α-particles. Explain how you would establish that it is only α-particles to which this detector responds. Also explain how the behaviour of this detector leads you to believe that the radiation detected is in the form of *particles*.

19-5 Describe a form of GM tube that can be used to detect α-, β-, and γ-radiation. Explain how you would use this tube to establish that all three kinds of radiation are present in the emissions from a radium source.

19-6 Under what circumstances would you expect the intensity of radiation from a small source to vary inversely wih the square of the distance from the source (i.e. so that doubling the distance reduces the intensity to a quarter as much)? Discuss to what extent you would expect such a variation to apply in the following cases:
(a) a light source in clear air
(b) an α-particle source in air
(c) a β-particle source (i) in a vacuum (ii) in air
(d) a γ-ray source in air.

Identifying the particles

19-7 Explain how you would establish the sign of the charge carried by β-particles.

19-8 In an experiment to measure the speed v and specific charge e/m of β-particles in a vacuum the following measurements are obtained:

$v/10^8$ m s^{-1}	1.5	2.0	2.5
$(e/m)/10^{11}$ C kg^{-1}	-1.54	-1.29	-0.96

$v/10^8$ m s^{-1}	2.7	2.9
$(e/m)/10^{11}$ C kg^{-1}	-0.78	-0.45

According to the special theory of relativity e/m should decrease with speed v according to the relation

$$\frac{e}{m} = \frac{e}{m_0} \sqrt{(1 - v^2/c^2)}$$

where m_0 is its rest mass and c is the speed of light. Plot a graph of e/m given in the above table against $\sqrt{(1 - v^2/c^2)}$. Use your graph (a) to decide whether the measurements confirm the special theory of relativity, and (b) to find the value of e/m at low speeds ($v \to 0$). Compare this value with the specific charge of cathode rays, and comment on the result.

19-9 Calculate the flux density of the magnetic field that will deflect into an arc of radius of curvature 0.10 m (a) β-particles of speed 2.0×10^7 m s^{-1} (b) α-particles of speed 2.0×10^7 m s^{-1} (whose specific charge is 4.8×10^7 C kg^{-1}).

19-10 Describe the differences that may be observed between the tracks of individual α-particles in cloud chambers filled with hydrogen, helium and air. Explain how information about the mass of an α-particle can be obtained from such data. Describe briefly an experiment with relatively large colliding bodies that supports your explanation.

19-11 Describe an experiment which suggests that most of the mass of an atom is concentrated in a small central region called a nucleus.

19-12 What is the wavelength of a γ-ray photon of energy 0.50 MeV?

19-13 A source containing radium-226 is found to emit γ-radiation of wavelength 6.5×10^{-12} m (as well as the usual α-radiation). What quantity of energy (in MeV) is carried away from the nucleus by each γ-ray photon?

Nuclear transformations

19-14 Samarium-147 (atomic number 62, symbol Sm) decays by α-emission. Explain what isotope it must decay into. The following is a list of neighbouring elements with their atomic numbers (in brackets): cerium (58), praseodymium (59), neodymium (60), promethium (61), europium (63), gadolinium (64).

19-15 A nucleus of radon-220 ($^{220}_{86}$Rn) decays by emission of an α-particle (a nucleus of $^4_2$He) of energy 6.3 MeV. Find
(a) the mass of the radon-220 atom
(b) the mass of the α-particle
(c) the nucleon number, atomic number, and mass of the resulting nucleus
(d) the speed and momentum of the α-particle
(e) the speed of recoil of the resulting nucleus
(f) the kinetic energy (in MeV) of the resulting nucleus.

19-16 (a) What energy (in J) is equivalent to a mass of 1.00 kg?
(b) Express this quantity of energy in eV.
(c) What is the kinetic energy (in eV) of an electron whose speed is such that its mass has increased to twice its rest mass?
(d) What p.d. is required to accelerate an electron to this speed?

19-17 Under certain circumstances γ-radiation can cause the production of pairs of electrons and positrons, the energy of each γ-ray photon being converted into the mass of an electron-positron pair. Taking the rest mass of an electron as equivalent to an energy of 0.512 MeV,
(a) what is the smallest energy (in eV) that a γ-ray photon must have in order give rise to an electron-positron pair
(b) what is the wavelength of such a γ-ray?
(c) If a γ-ray of half this wavelength produces an electron-positron pair and its energy is equally shared between the particles, what is the energy (in eV) of each of the particles produced?

19-18 Rubidium-87 ($^{87}_{37}$Rb) is a radioactive isotope that decays by emission of a negative β-particle. Write down the atomic number and nucleon number of the isotope into which it decays. If the relative atomic masses of the two isotopes are 86.909 186 and 86.908 892 respectively, calculate the maximum possible energy of the emitted β-particle (a) in J (b) in MeV.

19-19 An atom of phosphorus-32 decays by β-emission into an atom of sulphur-32, and in the process 1.7 MeV of nuclear energy is converted into kinetic energy of the particles. If the relative atomic mass of sulphur-32 is 31.972 1, calculate the relative atomic mass of phosphorus-32.

The neutron

19-20 Platinum-192 is a radioactive isotope that decays by emitting α-particles. State how many protons and how many neutrons there are both in the nucleus of platinum-192 and in the nucleus of the isotope into which it decays.

19-21 Potassium-40 (symbol $^{40}_{19}$K) is an isotope that decays by β⁺-emission. How many protons, neutrons and electrons are there in each neutral atom of this isotope? From the information in the following list (with atomic numbers given in brackets) write down the symbol of the isotope into which potassium-40 decays: chlorine (17), symbol Cl; argon (18), symbol Ar; calcium (20), symbol Ca; scandium (21), symbol Sc.

19-22 In a beam of slow neutrons uranium-238 ($^{238}_{92}$U) tends to absorb one neutron per nucleus, thereby turning into another isotope. This isotope decays by β-emission into neptunium (Np). The neptunium also decays by β-emission into plutonium (Pu). The plutonium is an α-emitting isotope. Write down the symbols (complete with supercripts and subscripts) for each of the isotopes involved, including the isotope into which the plutonium decays. Calculate the relative atomic mass of this isotope if the total kinetic energy of the particles emitted in the above processes is 11.05 MeV, the relative atomic masses of uranium-238 and helium-4 are 238.050 77 and 4.002 60, and the masses of the neutron and the electron are 1.008 67 m_u and 0.000 55 m_u.

19-23 The fission of one atom of uranium-235 releases 200 MeV of energy. A nuclear power station that uses uranium-235 has an output of 1.0 MW and is 40% efficient. Calculate (a) the number of atoms of uranium-235 that it uses per hour (b) the quantity of uranium-235 atoms (in mol) used per hour (c) the mass of uranium-235 used per hour.

19-24 In the interior of the Sun thermonuclear reactions take place whose net result is the conversion of hydrogen atoms into helium atoms with the release of energy. Calculate the energy released when four hydrogen atoms are converted into one helium atom, given that the relative atomic masses of hydrogen and helium are 1.007 82 and 4.002 60 respectively.

If it was possible to harness this reaction to provide the energy for a power station on Earth at an overall effieciency of 10%, estimate (a) the number of hydrogen atoms (b) the mass of hydrogen required per day to operate a 1000 MW power station.

Radioactive decay

19-25 Sketch a graph to show how the number of atoms in a radioactive sample varies with time and use it to show what is meant by the half-life of the substance.

A sample of a radioactive isotope with a half-life of 1.0 minute is obtained by chemical extraction from a nuclear reactor and deposited in a filter. If the sample contains initially 1000 atoms of the isotope, tabulate the number of atoms to be expected at subsequent intervals of 1 minute for the next 7 minutes, and represent your results graphically. Draw a tangent to the graph at the 2 minute mark and calculate the rate of decay at this moment. Calculate also the average decay rate between 1.5 min and 2.5 min from the start.

19-26 The graphs in the figure show how the numbers of atoms N in two radioactive samples A and B vary with time t (in days).
(a) Estimate the half-lives of the two samples, and calculate their ratio.
(b) At what time are there equal numbers of atoms in the two samples? Estimate the gradients of the two graphs at this moment, and calculate their ratio.
(c) With the aid of a ruler and set square estimate the moment at which the gradients of the two graphs are equal. Calculate the ratio of the numbers of atoms in the two samples at this moment.

From the above, suggest general rules (i) relating the activities of equal samples of radioactive materials with their half-lives (ii) relating the numbers of atoms in samples of equal activity with their half-lives.

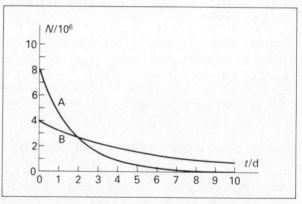

19-27 Technetium-99 (atomic number 43, chemical symbol Tc) decays by emission of a negative β-particle into ruthenium (chemical symbol Ru). Write down the full symbols with superscript and subscript for these two nuclei.

A sample containing 0.10 μg of technetium-99 is found to emit β-particles at a rate of 135 s⁻¹. Calculate
(a) the number of atoms N in the sample of technetium-99
(b) the decay constant $λ$ of technetium-99

(c) the half-life (in years) of technetium-99.

19-28 Calculate the mass of caesium-137 that has an activity of 5.0 μCi, given that the half-life of this isotope is 30 years.

19-29 Iodine-131 has a half-life of 8.0 days. A source containing this isotope has an initial activity of 2.0 Ci.
(a) What is the activity of the source after 24 days?
(b) What time elapses before the activity of the source falls to 1.0 μCi?

19-30 Describe how you would investigate the exponential decay law for a radioactive substance that can be placed inside an ionisation chamber. Explain how you would use your result to obtain the decay constant of the substance.

19-31 A neutron that is not bound in an atomic nucleus is found to be a radioactive particle with a half-life of 12.8 minutes. What is the probability of such a neutron decaying in a given interval of one second?

19-32 In order to find the volume of water in a central heating system a small quantity of a solution containing the radioactive isotope sodium-24 (half-life 15 hours) is mixed with the water in the system. The solution has an activity of 1.6×10^4 s^{-1}. When 30 hours have elapsed, it is assumed that the sodium-24 has mixed thoroughly with the water throughout the system, and a 100 ml sample of the water is drwn off and tested for radioactivity. It is estimated that the activity of the sample is 2.0 s^{-1}. What is the total volume of water in the central heating system?

19-33 A sample of a gaseous compound of uranium-235 is injected into the space inside a GM tube, and a count rate of 2000 per minute is obtained. In the absence of the uranium compound the background count rate was 100 per minute. If the sample of the gaseous compound contained 0.40 mg of uranium-235, calculate
(a) the amount (in mol) of uranium-235 atoms in the sample
(b) the number of uranium-235 atoms in the sample
(c) the activity of the sample
(d) the decay constant of uranium-235
(e) the half-life (in years) of uranium-235.

After the above measurements have been made it is found that the count rate increases over a period of hours, and when the measurements are repeated several days later the additional count rate caused by the uranium-235 has almost doubled. How do you account for this?

19-34 Without consulting any tables of radioactive series, calculate how many α-particles are emitted altogether by an atom of uranium-238 as it decays progressively into an atom of lead-206.

19-35 Uranium-234 is formed as one of the decay products of uranium-238. The half-lives of the two isotopes are 2.5×10^5 y and 4.5×10^9 y respectively.
(a) What proportion of the atoms in a sample of natural uranium would you expect to be uranium-234?
(b) What proportion of the mass of the sample would be in the form of uranium-234?

19-36 A compartment on a GM tube is filled with a solution containing 1.00 g of carbon extracted from an ancient document. The count rate recorded is 1000 per hour. When a similar solution containing 1.00 g of carbon extracted from a living plant is used instead, the count rate is 1200 per hour. Without any solution in the compartment the background count rate is found to be 300 per hour. Estimate the age of the document if the half-life of carbon-14 is 5730 years.

19-37 In a piece of living timber 1.25×10^{-12} of the total carbon content is in the form of the radioactive isotope carbon-14 (half-life 5730 years). A sample of carbon dioxide containing 2.00 g of carbon from living timber is introduced into a GM tube. Calculate
(a) the number of atoms of carbon-14 in the sample
(b) the decay constant of carbon-14
(c) the average number of disintegrations to be expected in 10 minutes, if the background count obtained with a non-radioactive gas in the GM tube is 10 per minute.

If the count rate obtained with an identical sample prepared from an ancient piece of timber is 307 in 10 minutes, calculate (d) the age of the ancient piece of timber.

20 Alternating currents

Data $c = 3.00 \times 10^8$ m s^{-1}

Measurements

20-1 An insulated wire of resistance 4.8 Ω is immersed with a thermometer in a quantity of oil held in a metal container; the heat capacity of this container and its contents is 450 J K^{-1}. Initially the oil is at a temperature below room temperature. An alternating current is then passed through the wire, and the temperature rises. At the moment when the oil is at the same temperature as the room the rate of rise is 4.0 K min^{-1}. Calculate the r.m.s. value of the alternating current.

20-2 The circuit shown in the figure is designed to provide a means of calibrating an a.c. ammeter to read r.m.s. values of current. Explain how you would use it and how the circuit achieves its purpose.

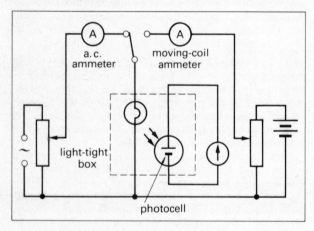

20-3 The r.m.s. value of a sinusoidal alternating current in a circuit is 2.0 A. What is its peak value? What is the peak value of the p.d. across a resistor of resistance 12 Ω in which this current flows?

20-4 A 12 V car battery and a low voltage mains supply are joined alternately to a car headlamp bulb, and both are found to keep it at the same brightness. What can you deduce from this about the mains supply? An oscilloscope is connected across the bulb while these tests are being conducted and the time-base is adjusted to give stationary traces. The Y-input switch is on 'direct', and the Y-amplifier control is set at 5 V div^{-1}. Describe what you would expect to observe on the screen in each case.

20-5 A television set is labelled at its mains input lead '240 V, 150 W, a.c. 50 Hz'. Write down an expression that shows how the current I taken by the set varies with time t, and state the values of any constants that appear in your expression.

What is the maximum power taken by the set at any instant?

20-6 An r.m.s. current of 13 A is being taken from the 50 Hz mains supply.
(a) What is the peak value of the current?
(b) What is the maximum rate of change of current?
(c) What is the current 300 μs after it changes direction?

20-7 A twin electric power cable for a 400 kV a.c. mains circuit has a total resistance of 16 Ω. The power supplied is 200 MW.
(a) How much does the potential difference between the two cables drop from one end to the other?
(b) What is the percentage loss of power as internal energy in the cable?
(c) What is the maximum p.d. for which the cable must be insulated?

20-8 A sinusoidal alternating potential difference is applied to the Y-amplifier of an oscilloscope, and the resulting trace is found to be 4.0 divisions in height from trough to crest. If the amplifier control is set at 5 V div^{-1}, calculate (a) the peak value of the alternating p.d. (b) the r.m.s. value of the p.d. (c) the r.m.s. value of the current through a resistance of 100 Ω to which this p.d. is applied.

29-9 What is meant by the *mean value* (or *half-cycle average*) of a sinusoidal alternating current I? Such a current is given by $I = I_0 \sin 2\pi f t$, where $I_0 = 1.0$ A and $f = 50$ Hz. Plot a graph of I against t, plotting points at 1 ms intervals from $t = 0$ to $t = 10$ ms.

The area under such a graph gives the charge that passes in the time interval considered.
(a) Estimate the total charge that flows in the circuit in one complete half-cycle.
(b) Calculate the half-cycle average current.

20-10 A square-wave alternating p.d. generated by a certain oscillator may be described as follows. For a time interval of 1.0 ms the p.d. is constant at its peak value $+V_0$; the p.d. then changes very rapidly to the value $-V_0$ at which it remains for a further 1.0 ms; it then reverts equally rapidly to the initial value $+V_0$, and so the cycle repeats continually. Sketch a graph of this waveform. What are the r.m.s. value and the half-cycle average value of this alternating p.d.? Also what is the frequency of this p.d.?

Phase differences

20-11 The figure shows two different phasor diagrams. In each there are two p.d.s represented V_1 and V_2. Draw

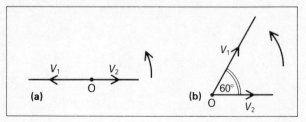

graphs of V_1 and V_2 against time in each case to show how these p.d.s are related.

20-12 In the figure the phasor OA rotates at constant angular speed in the usual way; but the end A, instead of moving in a circle, moves rather in a tightening spiral, as shown. Draw a graph of the p.d. V represented by this diagram.

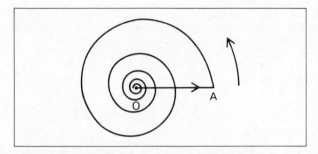

20-13 A rectangular coil is mounted in such a way that it can spin about an axis in its own plane at right-angles to a uniform magnetic field. The angle θ measures its rotation from the position $\theta = 0$ in which its plane is parallel to the field. What is the phase difference between the flux linked with the coil and the e.m.f. induced in it? Draw graphs one beneath the other to show how these two quantities vary with θ when the coil is rotating at a steady speed.

The peak value of the induced e.m.f. is found to be 50 V. Calculate (a) its value when $\theta = 60°$ (b) its r.m.s. value.

20-14 The alternating p.d. V across a mains supply is given by the equation $V = V_0 \sin 2\pi f t$, where $V_0 = 300$ V and $f = 50$ Hz. Plot a graph of V against time t, taking points at 1 ms intervals from $t = 0$ to $t = 20$ ms. Draw a tangent to the graph at the point where $t = 10$ ms, and work out the rate of change of V at this point.

A neon bulb lights only when the p.d. across it rises above 160 V, and it then stays lit until the p.d. drops below 110 V. If this bulb is connected across the above alternating supply, use your graph to calculate (a) for what period of time in each half-cycle the bulb is alight (b) for what percentage of the time the bulb is alight.

20-15 The figure shows a phasor diagram representing two alternating p.d.s of the same frequency across two components

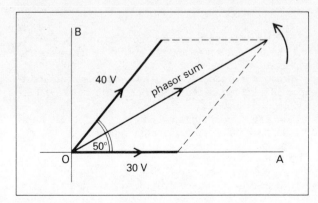

in series in a circuit. The p.d.s are of peak values 30 V and 40 V and the phase difference between them is 50°, as shown. The figure also shows the *phasor sum* of the two phasors, drawn by the same geometrical construction as for vector addition. Copy this figure on a larger scale on graph paper, and show that the sum of the components of the two p.d.s in the direction of the base line OA is equal to the component of the phasor sum in this direction. Now redraw the figure with the phasor system rotated (a) through 60° (b) through 120°, and show in each case that the same result applies.

Suggest a general rule for constructing the phasor representing the sum of two p.d.s of the same frequency.

20-16 Draw a complete circuit diagram of a three-phase power transmission system, showing the generator and a set of cables by which the power is taken to three separate housing areas (each of which you may represent on your diagram by a single resistor). Explain how it comes about that, if the same current is taken by each housing area, there is then no current in the neutral wire.

The neutral wire is earthed at the generator. Explain how there may nevertheless be a small potential difference between the neutral terminal and the earth terminal in a socket in one of the houses. Explain the function of the earth terminal as a safety device.

Reactive circuits

20-17 A sinusoidal alternating p.d. is applied across a capacitor. Explain how the current in the capacitor-leads varies with time, illustrating your explanation with free-hand graphs, one above the other, of p.d. and current against time.

Explain how the current is affected (a) by increasing the frequency (b) by increasing the capacitance.

20-18 A 16 μF capacitor is connected across the 240 V 50 Hz mains supply. Calculate the current in the wires leading to the capacitor. What p.d. must the insulation of the capcitor be able to withstand?

20-19 Calculate the largest capacitance that may be joined directly across the 240 V 50 Hz mains supply without risk of melting a 13 A fuse in the circuit.

20-20 A sinusoidal alternating current is driven through an inductor of negligible resistance. Draw graphs one under the other to show how the current through the inductor and the p.d. across it vary with time, and explain the relationship between the two curves.

For a given value of the p.d. explain the effect on the current of (a) increasing the frequency (b) increasing the inductance.

20-21 A choke (inductor) is to be used to limit the current in a fluorescent lighting tube to 2.0 A. Calculate the maximum value of inductance that could be needed for this purpose on a 240 V 50 Hz main supply.

..z A sinusoidal alternating current of peak value 50 mA .s flowing in a capacitor and an inductor joined in series. The peak values of the potential differences measured across each component are: capacitor 2.0 V, inductor 3.0 V. Draw accurately three graphs one under the other showing how the current and the two p.d.s vary with time. Use your graphs to give the sum of the p.d.s for a number of points in time, and plot a fourth graph showing the variation of the total p.d. across the combination. How do you describe this curve as to shape, frequency, magnitude and phase?

Write down the reactance of each of the components in this circuit, and deduce their combined reactance.

20-23 Explain what is meant by the *reactance* of a component. Write down the expressions you would use to calculate its value (a) for an inductance L, (b) for a capacitance C.

An unknown component is joined in series with an a.c. ammeter and a wattmeter to an alternating supply of r.m.s. value 10 V and frequency 1.0 kHz. The ammeter reads 0.25 A, but the wattmeter reads zero. Calculate (a) the reactance of the component (b) its inductance if it is an inductor (c) its capacitance if it is a capacitor. How, by varying slightly the frequency of the supply, could you distinguish whether the component is an inductor or a capacitor?

20-24 A coil of negligible resistance and with a laminated iron core is joined in series with a moving-iron ammeter across the 240 V 50 Hz mains supply. The ammeter reads 0.40 A. Calculate the inductance. Explain how it comes about that the average power drawn from the mains is zero, and give graphs of V, I and the product VI to show how the flow of energy into and out of the coil varies with time.

Mixed circuits

20-25 A coil has a resistance of 30 Ω and an inductance of 500 μH. Calculate its impedance (a) at 1.0 kHz (b) at 10 kHz (c) at 100 kHz. Comment on these values.

20-26 When a certain coil is joined across a lead-acid cell of e.m.f. 2.0 V, the current in it is 40 mA. But when the coil is joined across a 50 Hz alternating supply of e.m.f. 2.0 V, the current is 25 mA. Calculate the following properties of the coil: (a) its resistance (b) its impedance at 50 Hz (c) its reactance at 50 Hz (d) its inductance (e) the phase angle between p.d. and current at 50 Hz.

29-27 A coil of inductance 32 mH and resistance 12 Ω is connected across a 12 V 50 Hz a.c. supply. Calculate (a) the reactance of the coil (b) its impedance (c) the current in it (d) the rate of conversion of electrical energy to internal energy.

20-28 A capacitor of capacitance 5.0 μF is joined in series with a resistor of resistance 500 Ω across the 240 V 50 Hz mains supply. Calculate (a) the impedance of the combination (b) the current taken (c) the phase difference between current and p.d.

20-29 A capacitor of capacitance 0.00 μF and a resistor of resistance 16 kΩ are joined in series as in the figure to form a filter circuit. A sinusoidal alternating p.d. of r.m.s. value 1.0 V and of frequency 200 Hz is applied between A and C. Calculate (a) the reactance of the capacitor (b) the impedance of the combination of resistor and capacitor (c) the r.m.s. current through the combination (d) the r.m.s. potential difference between B and C (e) the r.m.s. potential difference between A and B.

Repeat the above calculation for the following frequencies of applied p.d.: (i) 40 Hz (ii) 1000 Hz. How would you describe the filtering properties of this arrangement?

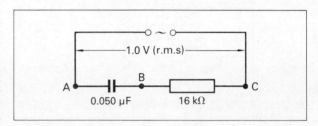

20-30 A resistor, a pure inductor, and a capacitor are joined in series with an ammeter of negligible impedance and connected to a 50.0 Hz supply of e.m.f. 12.0 V (r.m.s.), as shown in the figure. The r.m.s. values of the p.d.s across the three components are then measured, giving the figures shown. The ammeter reads 0.200 A. Draw a phasor diagram to show how these p.d.s are related and how they can be reconciled with the p.d. of 12.0 V across the combination of components. Calculate the resistance, inductance or capacitance of each of the components.

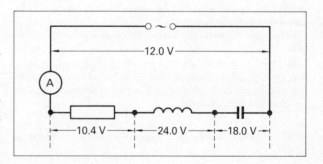

20-31 Sketch graphs one under the other to show how the p.d. across a capacitor and the current in its connecting wires vary with time.

A capacitor of capacitance 1.00 μF and a resistor of resistance 3.00 kΩ are joined in series across a 240 V 50.0 Hz a.c. supply. Calculate
(a) the impedance of the combination
(b) the peak value of the current through it
(c) the p.d. across the resistor at the instant when the current is a maximum
(d) the p.d. across the capacitor at the same instant
(e) the p.d. across the supply at the same instant.

Electrical oscillations

20-32 A coil (with inductance and resistance) is joined in parallel with a capacitor across a cell, as shown in the figure. When the current in the coil is steady, the switch is opened. Sketch a graph showing how the current in the coil then varies with time, and draw a second graph beneath showing the variation of the p.d. across the capacitor. Explain what is happening at different stages of the process, and what energy conversions are taking place.

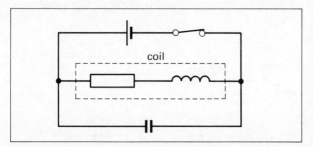

20-33 A teacher sets up the following mechanical arrangement as a model of an oscillating electrical system (of inductor, capacitor and resistor joined in a loop). He clamps one end of a steel strip in a vice in such a way that, when the strip is pulled to one side and released, it oscillates in a horizontal plane. At the other end of the strip he attaches a lump of metal sufficiently massive to slow down the oscillations to a readily observable rate. He explains that the horizontal displacement of the metal lump is equivalent to the p.d. across the capacitor, and sets the lump oscillating.

Explain how this model illustrates the behaviour of the electrical system; and point out what features of the mechanical system are equivalent to (a) the current in the circuit (b) the magnetic energy stored in the inductor (c) the electrical energy stored in the capacitor (d) the energy converted to internal energy in the resistor.

20-34 A coil of inductance 60 μH is to be connected with a capacitor to form a resonant circuit for a frequency of 1.2 MHz. What capacitance is needed?

20-35 A tuning circuit for a radio is to be constructed using a coil of inductance 465 μH and a variable capacitor with maximum and minimum capacitances of 208 pF and 10 pF. The stray capacitance of the rest of the wiring (effectively joined in parallel with the variable capacitor) is 10 pF. Calculate the maximum and minimum frequencies to which the circuit can be tuned.

20-36 What is the resonant frequency of a tuned circuit consisting of an inductance of 10 mH and a capacitance of 10 nF?

If these two components are joined in series and the coil is of resistance 100 Ω, sketch a phasor diagram showing the relations between the current and the p.d.s across capacitance, inductance and resistance. What is the impedance of the combination at the resonant frequency calculated above? If a p.d. of 1.0 V r.m.s. at this frequency is applied across the combination, what will be the p.d. across the capacitor?

20-37 A television signal is distributed to two buildings A and B by means of a coaxial cable which branches at a point X, one branch going to each building. When the cable is accidentally pulled out of the television set in building A, the set in building B shows a second faint image slightly displaced to the right. Explain how this comes about. If the second image is displaced 10 mm, and the speed of the spot of light across the screen is 7.0 km s^{-1}, estimate the length of the branch of cable to building A.

20-38 Calculate the length of a resonant dipole aerial for a radio transmitter at a frequency of (a) 30 kHz (b) 30 MHz (c) 30 GHz.

20-39 The relation for the speed c of electromagnetic waves:

$$c^2 = \frac{1}{\epsilon\mu}$$

applies not only to waves in a vacuum but also to waves travelling in any non-conducting medium.
(a) Calculate the speed of electromagnetic waves inside a ferrite block of relative permittivity 3.0 and relative permeability 100.
(b) Calculate the relative permittivity of glass at the frequency of light waves, given that the refractive index of glass for light is 1.50 and its relative permeability is 1.00.

21 Electronic devices

Junction diodes

21-1 Explain what is meant by *n*-type and *p*-type electrical conduction. How does the Hall effect provide us with a means of deciding which type of conduction is predominant in a slice of semiconductor? Explain with the aid of diagrams how you would know in a given case which type of conductor you were dealing with.

21-2 The figure shows a single crystal of silicon within which *n*-type and *p*-type regions have been formed as shown. The surface of the crystal has been covered with an insulating oxide layer through which a metallic contact is arranged to the top surface. Explain
(a) in what manner electrical conduction takes place in the *n*-type and *p*-type regions of the crystal
(b) how an electric field comes into existence across the junction between the *n*-type and *p*-type regions
(c) what happens when a p.d. is applied to the contacts making A positive with respect to B
(d) what happens when a p.d. is applied to the contacts making B positive with respect to A.

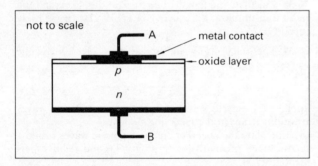

not to scale

21-3 The characteristic of a certain kind of silicon diode in the foward direction is shown in figure (a). (In the backward direction its resistance is effectively infinite.) This diode is joined in series with a resistor and milliammeter across a cell of e.m.f. 2.0 V; the current is found to be 4.0 mA.
(a) What is the p.d. across the diode?
(b) What is the p.d. across the resistor and milliammeter?
(c) Calculate the resistance of the resistor and milliammeter?
(d) If the cell is replaced by a variable supply, what p.d. must be used to give a current of 50 mA?

21-4 Considering the diode whose characteristic is given in figure (a) above, calculate the resistance of the diode (i) for an applied p.d. of 1.0 V (ii) for an applied p.d. of 0.75 V.

It is suggested that a silicon diode of this sort behaves when p.d.s are applied to it as though it consisted of a source of e.m.f. *E* in series with a resistance *R* and an idealised diode (whose resistance is zero in the forward direction and infinite in the reverse direction), as represented in figure (b)

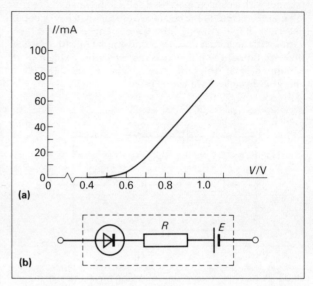

(a)

(b)

above. Draw a sketch-graph to show the characteristic of such a combination of components, and work out the values of *E* and *R* which approximate most closely to the characteristic of the actual diode.

21-5 A sinusoidal alternating supply of peak p.d. 2.5 V is joined across a silicon diode in series with a resistance of 24 Ω. The diode has a characteristic like that discussed in the previous two questions and represented by the idealised combination in figure (b) above.
(a) At what forward p.d. does the diode start to conduct?
(b) Calculate the peak value of the current.
(c) What is the peak value of the p.d. across the 24 Ω resistor?
Plot a graph showing how the p.d. of the supply varies with time. On the same axes plot a graph to show the variation of the p.d. across the 24 Ω resistor.

21-6 Two silicon diodes are connected in parallel as shown in the figure (a), and these are joined through a 560 Ω resistor to a sinusoidal alternating supply of peak value

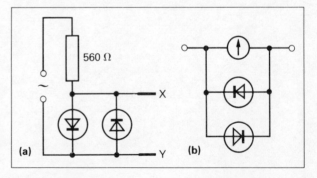

(a)

(b)

2.5 V. If the diodes have the same properties as that discussed in questions **21-3** and **21-4** above, draw two graphs on the same axes to show how (i) the supply p.d. and (ii) the p.d. between X and Y vary with time. How would you describe the waveform of the p.d. between X and Y?

21-7 The meter shown in figure (b) above has a resistance of 1000 Ω and gives full-scale deflection for a current of 100 μA. As a protective device it is supplied by the manufacturers with a pair of silicon diodes permanently connected across its coil, as shown. It is estimated that this should usually limit the current through the meter to about 6 times the full-scale deflection current. Explain how the device works.

21-8 Explain how *p-n* junctions are used in devices (a) to change light energy into electrical energy (as in a photo-diode) (b) to change electrical energy into light energy (as in a light-emitting diode). Explain in simple terms the physical principles involved in the operation of each device.

21-9 The figure shows the arrangement used in a dark-room exposure meter for an enlarger. The cadium sulphide photo-conductive cell is placed in the position where the photographic printing paper is to be exposed, choosing a part of the picture at which a mid-grey tone is wanted. The aperture controlling the amount of light passing through the enlarger lens and the variable resistor are then adjusted until the meter reads zero. The dial attached to the variable resistor knob is marked in *seconds*. Explain how the device works.

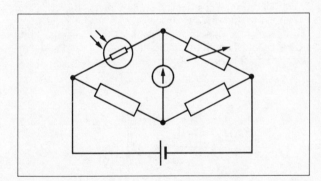

21-10 The detecting circuits used to rectify the small high-frequency alternating p.d.s in a radio set usually employ germanium diodes rather than silicon ones. Explain why this is so. For what applications would a silicon diode be preferable?

21-11 The figure shows a form of rectifier circuit employing a pair of diodes; it is supplied from a centre-tapped transformer winding, across the *whole* of which the r.m.s. potential difference is 12 V (at a frequency of 50 Hz). Describe how this circuit works, explaining what happens on successive half-cycles of the alternating supply. Sketch the waveform to be expected across the load resistor R_L.

Assuming that the drop in p.d. across the diodes in the forward direction is negligible, calculate

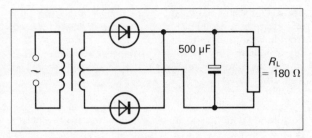

(a) the maximum p.d. across the load resistor R_L
(b) the frequency of the ripple across R_L
(c) the maximum current taken by the load resistor
(d) the amount the p.d. across R_L decreases during one cycle of ripple
(e) the average p.d. across R_L
(f) the maximum reverse p.d. across either diode.

Transistors

21-12 'The functioning of a transistor can be compared with the power-assisted brake system of a large vehicle, in which the power to work the brakes is drawn from the engine but is controlled by small movements of the brake pedal which require little power.'

Explain the meaning of this statement, and discuss the suitability of the analogy it draws between a transistor and a power-brake system. Draw a circuit diagram and explain the working of an amplifier employing a single transistor of some kind.

21-13 The magnetic relay in figure (a) operates at a current of 40 mA. The transistor has a current amplification factor of 80, and when it is conducting the p.d. V_{be} between base and emitter is 0.7 V. The resistance R of the photoconductive cell joined to its base varies with the amount of light falling upon it. Calculate the value of R at which the relay operates.

Suggest *two* possible practical applications of the circuit shown, and explain briefly how they work.

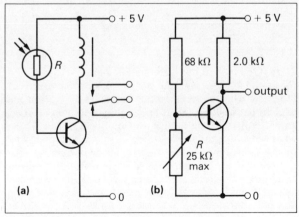

21-14 Describe and explain what happens in the transistor circuit in figure (b) above as the rheostat is slowly reduced from its maximum value to zero.
(a) Calculate the value of the resistance R of the rheostat at which the transistor just cuts off, if the p.d. V_{be} between base and emitter is 0.5 V at this point.

When the transistor is conducting V_{be} rises to 0.7 V and the current amplification factor is 100. If the transistor is just bottomed (potential at the output close to zero), calculate (b) the collector current, and (c) the base current.
(d) What is the p.d. across the 68 kΩ resistor and the current in it?
(e) What is the current in the rheostat and the p.d. across this?
Hence (f) calculate the value of R at which the transistor just bottoms.

Sketch a graph to show how the potential of the output varies with R.

21-15 Figure (a) shows part of a transistor amplifier. The transistor has a current amplification factor of 200, and in this circuit the p.d. V_{be} between base and emitter is always close to 0.7 V as along as the transistor is conducting. If no current is entering through the terminal A, calculate (a) the base current I_b (b) the collector current I_c (c) the collector potential V_c.
When a current I enters through the terminal A, calculate (d) the value of I for which the transistor bottoms, i.e. V_c falls to its minimum value of 0.3 V (e) the value of I for which the transistor is cut off, i.e. the current through it is zero and V_{be} falls to 0.5 V.

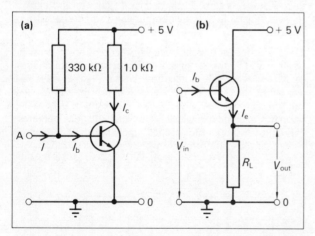

21-16 One type of transistor amplifier has the load resistor R_L joined in series with the *emitter*, as shown in figure (b). Describe what happens in this circuit as V_{in} varies, and explain (a) whether this is an inverting or a non-inverting amplifier (b) why the voltage gain must be less than 1, (c) in what sense this can be described as an amplifier, i.e. what quantites are *amplified* between input and output.
You can assume that for this transistor the current amplification factor is 50, and that the p.d. V_{be} between base and emitter is close to 0.7 V while the transistor is conducting, and only falls to 0.5 V when it is cut off. If $R_L = 430 \ \Omega$,

calculate (d) the values of V_{out}, I_e and i_b when $V_{in} = +5.0$ V (e) the value of V_{in} when $V_{out} = 0$ (f) the voltage gain of the amplifier.

21-17 The figure shows a device that may be used to supply a load R_L with a current I that remains constant for a large range of possible values of the resistance R_L. Explain how the circuit works, and calculate the current I for the values of p.d.s and resistances shown in the figure. You can assume that the p.d. V_{be} between base and emitter is 0.7 V. Estimate the maximum value of R_L for which the current I remains constant. What occurs if this value is exceeded?

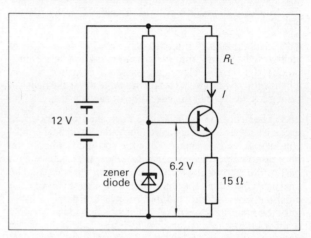

21-18 The figures shows three different forms of amplifier circuit based on i.c. operational amplifiers. In each case explain how the circuit works, and write down its gain. [Hint: remember that the potentials of the two inputs of the i.c. stay almost equal to one another as they vary, and that the currents into the inputs are always negligible.]

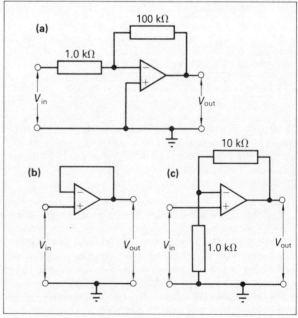

Sound reproduction

21-19 Describe two different physical processes in which a potential difference is produced by the mechanical vibrations caused by a sound wave.

21-20 A crystal pick-up used in a record player can produce a maximum p.d. (in either direction) of 150 mV. It is connected to the input of a transistor amplifier whose first stage has a voltage gain of −30. What is the minimum d.c. supply p.d. required to operate this stage? What is the significance of the minus sign used to state the voltage gain?

Switching circuits

21-21 Explain what occurs in the transistor circuit in figure (a) as the slider C of the potentiometer is slowly moved from position A to position B. Draw a sketch graph to show how the potential of the output of the circuit varies with the distance of C from A.

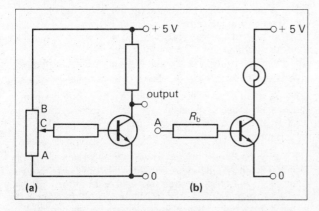

21-22 The arrangement in figure (b) may be used to indicate the state of the output of a switching circuit (which may take one of two values, either + 5 V or a value close to zero). The transistor has a current amplification factor of 120, and the p.d. V_{be} between base and emitter is 0.7 V when the transistor is conducting. The lamp takes a current of 60 mA when fully lit.
(a) What is the state of the output of the switching circuit indicated by the lamp being lit?
(b) What is the p.d. across the base resistor R_b when the input A is at a potential of 5 V?
(c) What is the base current when the lamp is fully lit?
(d) What resistance should therefore be selected for R_b?
(e) Using this value of R_b, if the lamp just lights weakly for a current of 40 mA, what is then the potential of the input A?

21-23 Explain the logical function performed by each gate in the figure (top right). In (a), (b) and (c) the object is to turn on the lamps with the switches concerned. In (d) and (e) the inputs A_1 and A_2 may take ether of the values 0 or + 5 V depending on the states of other circuits connected to them. The output of the gate is taken from B in similar form.

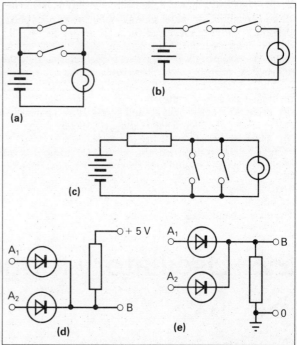

21-24 The figure below shows the design of a *decade counter*, i.e. a counter whose output B changes its potential from 5 V to 0 after 10 such changes of the input terminal A, at which point the circuit returns to its original state. The counter consists of four bistable circuits, each of which change their states every time their input terminals go from 5 V to 0. These are interconnected as shown with a 3-input OR gate and a 2-input OR gate; the output of an OR gate is at zero potential only when all its inputs are at zero potential. The initial state, represented in the figure, is with the right-hand transistor of each bistable conducting and all the lamps out, indicating the binary number 0000. Write down the sequence of binary numbers represented as successive input pulses are applied at A, and explain why the counter reverts to its initial state after the 10th pulse and does not proceed to show the binary number for ten.

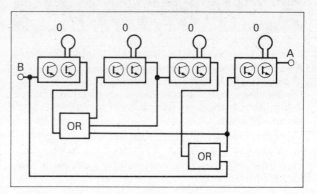

21-25 The device shown in the figure may be connected to a three-stage binary counter to indicate the number of pulses that the counter has received. Each of the inputs, A_1, A_2, A_3, is connected to the appropriate point in one of the bistable circuits of the counter; the inputs can thus take the values 5 V or nearly zero, depending on the state of the bistable circuit concerned. Calculate the p.d. registered by the high-resistance voltmeter

(a) when A_1 is at 5 V, but A_2 and A_3 are at zero potential

(b) when A_1 and A_2 are at 5 V, but A_3 is at zero potential

(c) when all three inputs are at a potential of 5 V.

Explain the general rule by which the reading of the voltmeter may be related to the number of pulses counted.

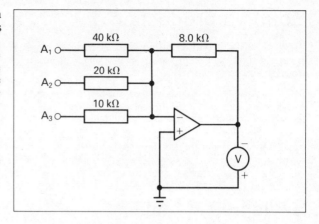

22 Oscillatory motion

Data $g = 9.81$ m s^{-1}

Describing oscillations

22-1 Sketch a displacement-time graph for a body moving with s.h.m. (a) Mark with an X points at which the speed of the body is a maximum. Is the velocity the same at all these points? (b) Mark with a Z points at which the speed of the body is zero.

22-2 The photograph shows a multiflash photograph of half an oscillation of a pendulum bob moving from left to right. the stroboscope was set to produce 30 flashes per second and the pendulum was 0.80 m long. Plot a graph to show how the horizontal displacements of the bob from the central position varies with time.

22-3 (a) Plot a velocity-time curve for the pendulum bob in the previous questions by drawing tangents at not less than six points on your displacement-time curve.

(b) Similarly plot an acceleration-time graph for the pendulum bob. Is there any simple relationship between the values of a and s?

22-4 Draw a velocity-time graph for two bounces of a bouncing rubber ball. Is this an example of s.h.m.? [Remember that the ball has a constant downward acceleration when in flight.]

22-5 What is (i) the period, and (ii) the frequency of

(a) the rise and fall of the sea

(b) the beat of a normal heart

(c) the oscillation of a ticker-timer arm

(d) the piano strings which oscillate when middle C is played? Express your answers in seconds and hertz respectively.

22-6 A small piece of cork in a ripple tank oscillates up and down as ripples pass it. If the ripples travel at 0.20 m s^{-1}, have a wavelength of 15 mm and an amplitude of 5.0 mm, what is the maximum speed of the cork?

22-7 A potential difference which alternates sinusoidally is applied to the Y-plates of a c.r.o. which has a calibrated time base. A stationary trace, with an amplitude of 4.0 div and a wavelength of 1.5 div, is obtained with the time base set at 1.0 ms div^{-1}. When the time base is switched off the trace becomes a vertical line. Calculate the maximum speed of the spot of light on the screen when producing the vertical line if 1 div is equal to a length of 10 mm.

22-8 The equation defining linear s.h.m. is
$$a = - \text{(constant)}s$$
What units must the constant have?
Two s.h.m.s A and B are similar except that the constant in A is ten times the constant in B. Describe how these s.h.m.s differ.

22-9 A body oscillates with s.h.m. described by the equation
$$y = (1.6 \text{ m})\sin(3\pi \text{ s}^{-1})t$$
What are (a) the amplitude and (b) the period of the motion? Calculate for $t = 1.5$ s (c) the displacement (d) the velocity and (e) the acceleration of the body. (f) If the body has a mass of 0.80 kg what is the resultant force on it at time $t = 1.5$ s?

22-10 The figure shows a body which is moving in a circle of radius 1.2 m at 100 revolutions per minute. Calculate the period of its circular motion and deduce the time interval between adjacent positions AB, BC, etc.

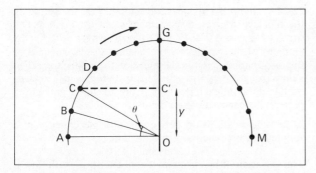

By drawing the figure to scale or by calculation produce a set of corresponding values for θ and y, where θ is measured clockwise from OA and y is the distance from the centre of the circle O to the foot of the perpendicular C′ from C to OG. Plot a graph of y against θ. Use the figure and your graph to explain the relationship between uniform circular motion and s.h.m.

22-11 A dock has a tidal entrance at which the water is 10 m deep at 12 noon, when the tide is at its lowest. The water is 30 m deep when the tide is at its highest, which follows next at 6.15 p.m. A tanker, needing a depth of 15 m, requires to enter the dock as soon as possible that afternoon. Calculate the earliest time it could just clear the dock entrance. (Assume that the water rises and falls with s.h.m.)

Simple harmonic oscillators

22-12 The period T of vertical oscillations of a mass m supported by a light, helical spring of stiffness k (force per unit extension) is given by $T = 2\pi m/k$.

Two identical springs of force constant k are connected (a) in series (b) in parallel, and support a mass m. What is the ratio of the period of vertical oscillation of the series arrangement with that of the parallel arrangement?

22-13 A mass m oscillates with s.h.m. of period T_1 when hung from a spring of stiffness k. The spring is cut in half and when hung from one of the halves the mass oscillates with period T_2. Calculate the ratio T_1/T_2.

The mass is now hung from the two halves arranged in parallel. If the new period of oscillation is T_3, calculate the ratio T_1/T_3.

22-14 A body of mass 200 g is executing simple harmonic motion with an amplitude of 20 mm. The maximum force which acts upon it is 0.064 N. Calculate
(a) its maximum acceleration
(b) its period of oscillation
(c) its maximum speed.

22-15 The piston of a car engine has a mass of 0.40 kg. It moves with simple harmonic motion of amplitude 40 mm. Explain at what point(s) in its motion the piston experiences a maximum force F_o and has maximum kinetic energy E_o. When the engine is rotating at 5000 rev min^{-1} calculate the values of F_o and E_o.

22-16 Find the length of a 'seconds-pendulum', that is a simple pendulum with a period of 2.000 s, at a place where $g = 9.182$ m s^{-2} (the value at the National Physical Laboratory at Teddington, England).

22-17 From the multiflash photograph of question **22-2**, measure the time period T of the simple pendulum and calculate $gT^2/4\pi^2$. How does your value compare with the length of the pendulum given in the question?

22-18 Two simple pendulums are suspended side by side from two fixed supports in the same horizontal plane. They are displaced by equal amounts and released simultaneously so that they oscillate, initially in phase with one another, in two parallel vertical planes. If the two pendulums differ in length by a small amount, describe their relative motions when viewed in a direction perpendicular to the vertical planes in which they oscillate. If the lengths of the two pendulums are 0.490 m and 0.340 m, calculate the number of oscillations made by each pendulum between successive times when they are exactly in phase. Explain your method of calculation.

22-19 In the figure, (a) shows a loaded tube floating in equilibrium in a liquid with a length l_o immersed and a free-body diagram for the tube. The tube is then pushed down and released; (b) shows the tube at the instant when its displacement is s from the equilibrium position together with the corresponding free-body diagram. By applying Newton's first law in (a) and his second law in (b) show that
$$ma = U - U_o$$
and hence, using Archimedes's principle, prove that the tube oscillates with s.h.m. State any assumption that you make.

Calculate the period of oscillation for a tube of mass 25 kg and cross-sectional area 0.040 m^2 in water of density 1000 kg m^{-3}.

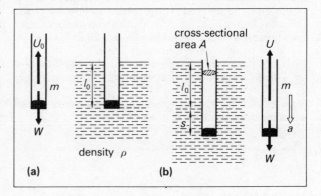

(a)   (b)

Oscillators and energy

22-20 A simple pendulum is set oscillating with an amplitude of 10°. What will be the amplitude of its oscillation when half of its initial energy has been converted to internal energy?

22-21 The figure shows a 'time-trace' made by a fine brush atached to the end of a vibrating arm while moving a piece of paper steadily past it. Draw sketch graphs to show (a) how the period T of the arm varies with the amplitude y_o of its vibration (b) how the amplitude y_o varies with the time t. [Measure the ratio of successive maximum displacements.] (c) How does the total energy E of the vibrating arm vary with time t?

Sketch the experimental arrangement used to produce the time-trace and describe how you would demonstrate the effect of increased damping on the system.

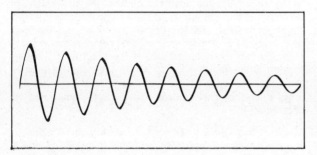

22-22 Give a diagrammatic description of the energy changes which occur in an oscillating system which consists of a metal sphere on one end of a long spring attached to a firm support, after the sphere is pulled down and released so that it moves in a vertical line with s.h.m. Assume that a negligible amount of energy is converted to internal energy during one complete oscillation.

22-23 A block of wood mass 0.25 kg is attached to one end of a spring of constant stiffness 100 N m^{-1}. The block can oscillate horizontally on a frictionless surface the other end of the spring being fixed. The graph shows how the elastic potential energy E_p of the system varies with displacement s for a horizontal oscillation of amplitude 0.20 m.
(a) Show that the graph is correctly drawn.
(b) Copy the graph and sketch a second curve to show how the kinetic energy of the mass varies with the displacement s.
(c) Calculate the maximum speed of the block and its speed when the displacement is 0.10 m.

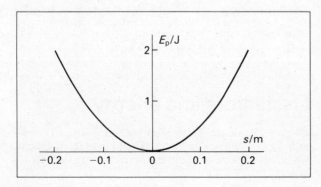

22-24 In the previous question the pull F of the spring on the block at a given displacement s can be found in two ways: (a) by using $F = ks$ and (b) by using $F = \Delta E_p/\Delta s$, the gradient of the graph of e.p.e. against displacement. Use the information given in the question to find the force F at values of s given by $s/m = 0, 0.05, 0.10, 0.15$ and 0.20 using both methods at each position. [Copy the figure onto graph paper before measuring gradients.]

22-25 The relationship between the displacement s of a simple harmonic oscillator and the time t is
$$s = (1.2 \text{ m}) \sin(4.0 \text{ s}^{-1})t$$
The variation of its elastic potential energy E_p with time is given by $E = \frac{1}{2} ks^2$, so that
$$E_p = (58 \text{ J})[\sin (4.0 \text{ s}^{-1})t]^2$$
when k, the restoring force per unit displacement of the oscilator, is 80 N m^{-1}.

Plot a graph of y against t and below it, using the same scale on the time axis, a graph of E_p^2 against t. [Be careful about the shape of the $E_p - t$ graph near points where $y = 0$.]

22-26 What is meant by critical damping? Describe *two* situations where it is desirable to have critical damping in a mechanical system and discuss what would be the result of (a) very heavy damping and (b) very light damping on each.

22-27 A punchbag of mass 0.65 kg is struck and oscillates with s.h.m. If the total mechanical energy of the oscillations is initially 55 J, what is the maximum speed of the punchbag? Describe how the energy of the punchbag changes as its oscillations die away.

22-28 If, in the previous question, the frequency of oscillation of the punchbag is 2.8 Hz, calculate (a) the amplitude of its oscillation, and (b) the stiffness, i.e. the force per unit extension, of the supporting spring system.

Resonance

22-29 Describe how you would demonstrate that the sharpness of resonance in an oscillating system depends on the amount of damping which it experiences.

22-30 The figure shows a steel spring A, clamped horizontally at one end B, and loaded at the other with a small iron disc M. The natural frequency of the loaded spring is 20 Hz and it is driven by an electromagnet with poles P and Q connected to an a.c. generator. Describe the change in the amplitude of M as the frequency of the generator is increased slowly from 5 Hz to 50 Hz.

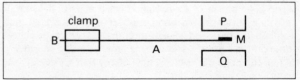

22-31 Here are a list of oscillating objects:
(a) a tuning fork prong
(b) a boat anchored in a swell
(c) a loudspeaker transmitting a pop-song
(d) a swinging leg while walking
(e) a singing wine glass.
Use the language of mechanical oscillations to describe and classify the objects on the list.

22-32 Explain the meaning of the terms (a) *damped oscillation*, (b) *forced oscillation*, and (c) *resonance*. Describe carefully one example of resonance in each of (i) mechanics, (ii) sound, and (iii) electricity.

22-33 A drilling machine was found to vibrate in such a way as to make accurate work impossible at certain frequencies. An investigation of its behaviour showed that the amplitude of the vibration of the drill bit s_o was related to its frequency of rotation f as follows:

$s_o/10^{-2}$ mm	8	14	30	44	80	96	
f/Hz	0	5	8	9	10	11	
$s_o/10^{-2}$ mm	24	8	2	3	9	7	4
f/Hz	12	13	15	20	25	30	35

Draw a graph of s_o (up) against f (along) and explain its shape. Why might it not be advisable to start to drill a hole with the drill rotating at a frequency of between 15 Hz and 20 Hz?

22-34 At certain definite engine speeds parts of a car, such as a door panel, may vibrate strongly. Explain the physical reason for this strong vibration and suggest how it might be reduced by alterations to (a) the engine assembly and (b) the door.

22-35 A loudspeaker is found to boom at a frequency of 75 Hz. How would you (a) lower the pitch at which booming occurs (b) making the booming less?

23 Describing waves

Data $g = 9.81$ m s^{-1} = 9.81 N kg^{-1}

What are waves?

23-1 The figure shows an idealised wave pulse travelling to the right along a heavy rope at 5.0 m s^{-1}.
(a) Sketch a graph to show how the displacement y_P at P varies with time t during the next 0.50 s.
(b) Add a second graph using the same time axis to show how the displacement y_Q of Q varies with t.
(c) How would your graphs differ if the wave pulse was travelling at 1.0 m s^{-1}?

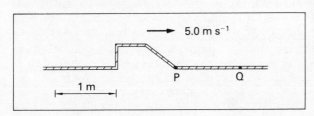

23-2 From the photograph plot the displacement y (positive to right) against position x for the spheres. Take the first and last sphere to be undisplaced and the remainder to have originally been evenly spaced between them. Draw a smooth curve through your points and comment on the result.

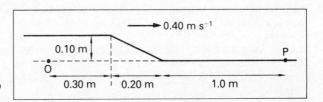

23-3 The figure shows an idealised wave pulse, a 'knee' travelling to the right along a long spring.
(a) Draw a graph to show how the *velocity* of the point P varies with time t, mark scales on your graph axes.
(b) Draw a second graph to show the transverse velocity of the spring against distance x from O at the instant shown in the diagram. Mark scales on your graph axes.
(c) How would your graph in (b) differ if the wave pulse was travelling at 2.0 m s^{-1}?

23-4 The photograph shows a wave pulse on a rope moving to the right. During the exposure the transverse movement of the rope has produced a blurring at some points. Draw a sketch of the pulse and indicate on it which parts of the rope are moving and in what direction. Explain how the energy of the wave pulse is stored at different places.

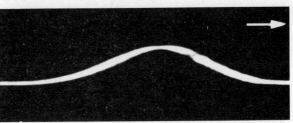

Wave speeds

23-5 A transverse wave is seen to travel along a length of cotton thread stretched to a tension of 1.5 N at a speed of 60 m s^{-1}. What is the mass of 100 m of the cotton thread?

23-6 A heavy chain is hanging vertically with its lower end free. Describe the motion of a wave pulse which travels down the chain. If the chain is 8.0 m long and has a mass of 24 kg calculate the speed c of the pulse at distances d from the top given by $d/m = 0, 2, 4, 6, 7, 7.5$. Hence sketch a graph of c against d.

23-7 The speed c of longitudinal waves on a stretched spring of large diameter (a 'slinky') is given by $c = \sqrt{(kl/\mu)}$, where k is the spring constant, l the stretched length and μ the stretched mass per unit length of the slinky.
(a) Show that this expression for c is dimensionally correct.
(b) Calculate c for a slinky of mass 0.45 kg which stretches to be 3.0 m long when the tension in it is 6.0 N. (Assume the unstretched slinky has a negligible length.)
(c) Show that the time for a longitudinal wave pulse to travel from one end of the slinky to the other is independent of its stretched length. Make the same assumption as in (b).

23-8 A steel ball of diameter 62 mm is bounced on a massive steel plate. The time of contact is found electrically to be equivalent to a trace of length 2.4 div on an oscilloscope screen with a time base set at 10 μs div^{-1}. Assuming that the time of contact is equal to the time taken for the longtitudinal wave pulse to travel from the 'bottom' of the steel ball to the 'top' and back again, calculate the speed of the wave pulse in steel.

23-9 For each of the following types of mechanical wave list their speed or range of speeds and their range of wavelengths:
(a) sound waves in air (b) seismic waves in the Earth's crust (c) surface waves on water.

23-10 Suppose that the speed of propagation c of longitudinal waves in a metal rod depends only on E the Young modulus of the material, ρ the density of the material and d the diameter of the rod. Work out a possible expression relating c to E, ρ and d and find a value for the constant given that for copper rod at a speed of 3570 m s^{-1} at 20°C. If the separation 8900 kg m^{-3}, and $c = 3800$ m s^{-1}, all to two significant figures.

23-11 A longtitudinal wave pulse is found to travel along a copper rod at a speed of 3570 m s^{-1} at 20°C. If the separation of copper atoms is 2.5×10^{-10} m and the mass of one copper atom is 1.06×10^{-25} kg, calculate the intermolecular spring constant k for copper. What assumption(s) do you make in your calculation?

23-12 A recording station observes that there is an interval of 68 s between the reception of P (push or primary) and S (shake or secondary) waves from an underground nuclear test explosion. If the speed of the P and S waves in the Earth's crust are 7800 m s^{-1} and 4200 m s^{-1} respectively, find the distance of the test site from the recording station.

23-13 The graphs in the figure show the variation of the speed c of water ripples (such as you might see in a laboratory ripple tank) with the depth of the water h.
(a) Calculate the times taken for ripples of frequency (i) 5.0 Hz and (ii) 15 Hz to transverse a tank 0.80 m long when the water is firstly 20 mm deep and secondly 2.0 mm deep.
(b) Sketch a graph of wave speed c against depth of water h for ripples of frequency (i) 5.0 Hz and (ii) 15 Hz.

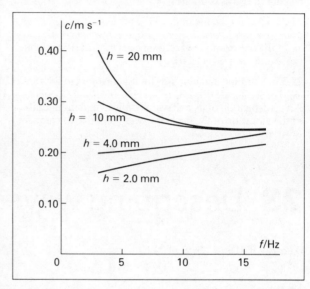

23-14 Assuming that the speed c of ripples of small wavelength λ depends only on λ, on the surface tension of the liquid γ and on the density of the liquid ρ, use the method of dimensions to deduce a possible expression for c.

The principle of superposition

23-15 What is meant by the principle of superposition of waves? Are there any types of wave which do not obey the principle or any circumstances in which it does not apply?

23-16 The figure shows two wave pulses at time $t = 0$. The markings on the x-axis are 1.0 m apart.
(a) Draw a series of y–x graphs at times t given by $t/s = 1, 2, 3, 4, 5$ to illustrate their superposition as they cross. [Hint: sketch the two pulses in lightly and then add the displacements to find the resultant pulse.]
(b) Draw two y–t graphs, one for the point P and one for Q, during the time $t = 0$ to $t = 8$ s.

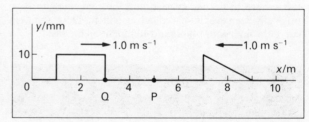

23-17 The figure shows two wave pulses at $t = 0$. The markings on the x-axis are 1.0 m apart.
(a) Draw a series of y–x graphs at times t given by $t/s = 0.05, 0.10, 0.15, 0.20, 0.25, 0.30$ to illustrate their superposition as they cross.
(b) Draw a y–t graph for the point P from time $t = 0$ to $t = 0.30$ s.

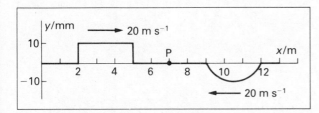

23-18 A wave pulse, similar in shape to that shown moving to the left in the figure accompanying the previous question is moving along a heavy rope which is attached to a fixed post. Sketch a series of diagrams to illustrate the reflection of the wave pulse at the post and explain in what way the principle of superposition helped you to draw the shape of the rope *during* the reflection.

Where is the energy of the wave pulse at the instant when the rope is entirely horizontal?

Repetitive waves

23-19 The figure shows a y–x graph for a wave moving to the right and about to pass through a point P. Draw the corresponding y–t graphs for P with marked scales on both axes. What is (a) the wavelength (b) the period, and (c) the frequency of this repetitive wave?

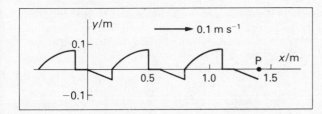

23-20 Two seagulls are observed to be bobbing up and down in antiphase as water waves pass them. The frequency of their oscillation is 0.40 Hz, they are 15 m apart and the wave speed is 4.0 m s^{-1}. Are these values compatible with the observation?

23-21 The figure shows a sinusoidal wave travelling to the right along a rope. The dark line represents the wave at $t = 0$ and the dashed line represents the wave at $t = 0.25$ s. What is (a) the amplitude y_0 and (b) the wavelength λ of this wave? Calculate (c) the speed c (d) the frequency f and (e) the period T of the wave.

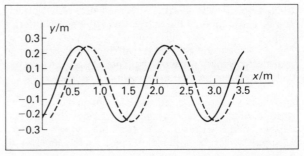

23-22 A photograph of a sinusoidal transverse wave on a string can be described by the equation
$$y = (0.02 \text{ m}) \sin (10 \text{ m}^{-1}) x$$
What is (a) the wavelength and, if the speed of the wave was 25 m s^{-1} (b) the frequency of the wave? (c) What is the maximum transverse speed of a point on the string as the wave passes?

23-23 A sinusoidal wave transmits a power of 4.0 W. Find the power it can transmit if (a) its amplitude is doubled, the frequency remaining constant, or (b) its frequency is trebled, the amplitude being halved.

23-24 The displacement y at time t produced by a wave travelling to the right along a rope may be represented by
$$y = a \sin \frac{2\pi}{\lambda} (ct - x)$$
where λ is the wavelength and c is the speed of the wave. What does a represent?
If $a = 0.20$ mm, $c = 9.0$ m s^{-1} and the wave has a frequency of 3.0 Hz, draw to scale (a) a graph of displacement y against distance from origin x at time $t = 0$, and (b) a graph of displacement y against time t for the point $x = 6.0$ m.

23-25 What is the result of superposing two sinusoidal waves of the same frequency, one of which has twice the amplitude of the other, at a place where they arrive with a phase difference of 90° ($\pi/2$ rad)? [Draw two y–t graphs on the same axis for the waves to be added and find the resulting y–t graph using the principle of superposition.]

Stationary waves

23-26 A demonstrator wishes to use a vibrator and a stretched rubber cord to show the production of stationary waves. If the cord has a mass of 0.12 kg and when stretched to a length of 2.4 m has a tension of 6.0 N in it, at what frequencies of vibration will the cord settle into stationary wave patterns?

23-27 A violin string of mass per unit 3.75×10^{-4} kg m^{-1} is stretched to a tension of 15 N. What is the fundamental frequency of a note played on this string when its length is restricted to (a) 0.30 m (b) 0.20 m (c) 0.15 m? What harmonics (overtones) may be present when the fundamental note is played on the 0.30 m length of string?

23-28 Explain why a string stretched between two fixed points can support stationary waves of many different frequencies, and why the frequencies form a sequence f_o, $2f_o$, $3f_o$, $4f_o$, etc. How would you attempt (a) to excite on its own the stationary wave of frequency $2f_o$, and (b) to completely suppress the stationary wave of frequency $2f_o$?

23-29 Draw successive positions in the vibration of a stretched string oscillating in its third harmonic. Hence explain why adjacent nodes in the stationary (standing) wave pattern are $\lambda/2$ apart, where λ is the wavelength of the superposing waves which are producing the stationary wave.

23-30 An aluminium rod 1.20 m long is suspended horizontally from two threads. When one end of the rod is tapped by a hammer the rod 'rings' emitting a note of frequency 2130 Hz.
Explain how the note is produced and calculate a value for the speed of longitudinal mechanical waves (compression waves) in the aluminium rod.

23-31 A sonometer wire (a wire stretched above a sounding box) has a fixed mass per unit length μ. The tension T in the wire and the length l of the wire can be independently varied. An experimenter measures the frequency f of the fundamental vibration of the wire under various conditions of t and l. Draw sketches to show the form of graph he gets when he plots
(a) f against T, keeping l constant
(b) f^2 against T, keeping l constant
(c) f against l, keeping T constant
(d) f^{-1} against l, keeping T constant.
Explain the theory behind your sketches and state, for any of the graphs which is linear, what is represented by its gradient.

23-32 If in the previous question sonometer wires of the same length but with different mass per unit length μ were all stretched to produce the same fundamental frequency how would the tension T vary with μ?

23-33 The lowest note on a piano has a frequency of 27.5 Hz. The wire which produces this note when struck is 2.00 m long and has a tension of 320 N. What is the total mass of the wire?

23-34 A wire of mass per unit length 1.6 g m^{-1}, is attached to a fixed block A and pulled by a spring balance B. The stretched piece of wire is 1.8 m long and its centre is placed between the poles of a large magnet. When an alternating p.d. of frequency 50 Hz is connected across the wire
(a) describe qualitatively what happens as the tension in the wire is slowly increased. [Hint: consider the possible motion of the centre of the wire.]
(b) Calculate the tension in the wire when it oscillates in its fundamental mode.

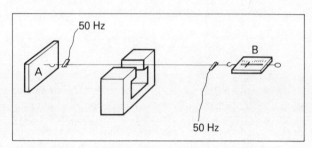

23-35 Suppose in the previous question the amplitude of oscillation of the centre of the wire in its fundamental mode is 12 mm.
(a) What is the maximum speed of this part of the wire?
(b) If the B-field between the poles of the magnet is 1.0×10^{-2} T, over what length of wire must this field extend if the peak value of the induced e.m.f. in the wire is to be 2 mV.

23-36 When the table on which a cup of tea or coffee is standing is knocked a stationary wave pattern often appears on the liquid surface for a short time. Sketch the form of the pattern and explain how it is produced.

23-37 Take a length of rope or chain or string and oscillate one end sinusoidally from side to side so that the freely hanging part forms a stationary wave pattern. What modes of vibration can you produce? Give a qualitiative explanation of the simpler modes. [See queston **23-6**.]

24 Sound

Data speed of sound in air = 340 m s^{-1}
speed of sound in water = 1500 m s^{-1}

The nature of sound

24-1 A steady note from a works siren is heard by two men, P and Q, who are standing in an open field; P is 2.0 m nearer to the siren then Q. If the frequency of the note is 850 Hz, what is the phase difference between the waves at P and Q?

24-2 Draw a single energy flow diagram to represent the action of a device which could act both as (a) a loudspeaker and (b) a microphone. Explain the action of one of these instruments.

24-3 Suggest how you could make a diffraction grating for ultrasonic waves of frequency 40 kHz. Give details of overall size, spacing materials used, etc.

24-4 Sketch displacement-time graphs to show the result of superposing two sound waves of frequency 500 Hz and 600 Hz which arrive with equal amplitudes at a point. (The t-axis should cover a time of about 0.03 s.)

Describe how a cathode ray oscilloscope and associated apparatus could be used to demonstate your resultant graph.

24-5 Stretch a piece of string about 1 m long between two fixed points. Hang two simple pendulums of equal length symmetrically from different points on the string. Set *one* pendulum swinging perpendicular to the stretched string. Now describe in the language of physics what subsequently occurs.

24-6 The diagram shows an experimental arrangement for investigating the superposition of sound from two sources S_1 and S_2. If the wavelength of the waves from S_1 and S_2 is 80 mm in air and the distance $S_1M = 0.80$ m, suggest three possible values for S_2M when (a) the height of the trace on the c.r.o. is a maximum (b) the height of the trace on the c.r.o. is a minimum.

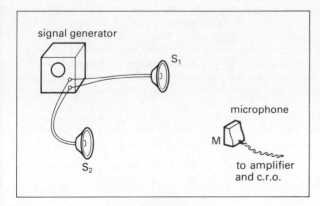

signal generator

S_1

microphone

M

S_2

to amplifier and c.r.o.

24-7 In another experiment with the apparatus described in the previous question the height of the trace on the c.r.o. is found to be a minimum when $S_1M = 0.80$ m and $S_2M = 1.00$ m. Why can the wavelength λ of the sound waves in air not be found from these measurements alone? What further observations are needed before λ can be found?

24-8 The diagram shows a simple apparatus whereby sound from a source at S can reach a listener with his ear held at E by two routes. When the slider is moved in or out the sound heard by the listener varies from maximum to minimum and so on. Explain this phenomenon.

With a source of frequency 1000 Hz it is found that the slider has to be moved out 160 mm between adjacent minima. Calculate the speed of sound in the tube.

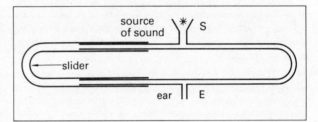

source of sound

S

slider

ear

E

24-9 Using the apparatus described in the previous question it is found that, with the slider fixed, variations of intensity are heard by the listener as the *frequency of the source* is varied. Explain this and support your explanation by giving some possible values for the quantities involves. [Take measurements from the diagram assuming that the length of overlap of the slider shown is about 0.5 m.]

24-10 A microphone with a circular diaphragm of effective diameter 23 mm is receiving a sound wave of intensity 1.2 μW m^{-2} (about the intensity of sound in normal conversation). How much energy does the microphone receive during 5 s?

24-11 A tuning fork of frequency 440 Hz is sounded together with a stretched metal wire. When the tension in the wire is 100 N, the experimenter hears exactly 2 beats per second. When he gradually reduces the tension in the wire without altering its length the beat frequency decreases to zero and then increases again to exactly 2 beats per second. Calculate the value of the new tension in the wire.

The speed of sound

24-12 What is the critical angle for sound waves refracted from air to water? Use your answer to explain why it is so quiet under-water. [See data.]

24-13 Dolphins communicate by emitting ultrasonic waves of frequency in the range 100 kHz to 250 kHz. What wavelength range in water does this represent?

24-14 A source of sound waves of frequency 570 Hz emits waves of wavelength 600 mm in air at 20 °C. Calculate the speed of sound in air at this temperature? What would be the wavelength of the sound from this source in air at 0 °C?

24-15 A man hums a note of frequency 250 Hz. He expels as much of the air in his lungs as he can and takes a deep 'breath' of helium. Estimate the note he now produces when he tries to hum at 250 Hz. Explain any assumptions you make in your answer. (Take $\rho_{air} = 1.3$ kg m^{-3}, $\rho_{He} = 0.18$ kg m^{-3}, $\gamma_{air} = 1.4$ and $\gamma_{He} = 1.6$.)

24-16 It is suggested that the speed of sound c in free air depends upon the air pressure p, the density ρ of the air and the wavelength λ of the sound. Use units or dimensions to predict the form of a possible relation between c, p, ρ and λ. What important property of sound does your answer imply?

24-17 On an open field an experimenter A bangs two pieces of wood together producing a regular series of sharp claps. A second experimenter B walks away from A and notices that when he has moved 120 m he hears the claps half way in time between seeing them. Calculate the time interval between the claps.

24-18 An object is vibrating vertically in a water surface with a frequency of 10 Hz. At the same time it is moving in a horizontal straight line with a speed of 20 mm s^{-1}. The waves it produces travel with a speed of 120 mm s^{-1}. Draw a diagram showing the instantaneous positions of the waves emitted during the previous half second. Calculate the wavelength of the waves (a) ahead of (b) behind the moving object.

24-19 A car travels with uniform speed along a long straight road (represented by the horizontal axis in the figure) sounding its horn continuously. A microphone connected to a frequency meter is placed close to the side of the road. It is found that the frequency f recorded by the meter varies with the position x of the car as shown. Explain the graph.

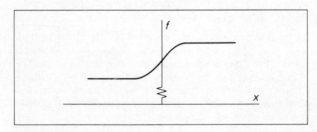

24-20 An observer accelerates from a speed of 170 m s^{-1} to a speed of 400 m s^{-1} while moving away from a source of sound of frequency 500 Hz. Describe what he would hear.

24-21 If the source of waves in question **24-18** moved at 200 mm s^{-1}, draw a diagram showing the instantaneous positions of the waves emitted during the previous half second. Calculate the angle of the cone formed by the 'sides' of the wavefronts which look like the familiar bow wave produced by a boat.

Vibrating air columns

24-22 A small loudspeaker emitting a note at 600 Hz is held above a tall glass measuring cylinder which is 0.50 m tall. Sketch a graph of the sound intensity I heard by a nearby observer against the depth of the water d as water is poured down the side of the measuring cylinder until it is full. Mark a rough scale on the d axis.

24-23 Two loudspeakers face each other at a separation of about 30 m. They are connected to the same sinusoidal oscillator which is set at 170 Hz. Describe and explain the variation of sound intensity heard by a man who walks at a slow steady speed of 0.50 m s^{-1} along the line between the two speakers.

24-24 Three identical springs A, B and C are each set vibrating in longitudinal stationary wave modes. Spring A has two fixed ends, spring B has two free ends and spring C has one end fixed and one end free.
(a) What frequencies in terms of f_A, f_B and f_C (the fundamental frequencies of A, B and C) are possible for each of A, B and C?
(b) What are the values of f_B and f_C if $f_A = 30$ Hz?

24-25 The air in a resonance tube closed at one end is made to vibrate by a small loudspeaker placed over the other end. The length l of the tube can be varied. The lowest frequency f which produces resonance for various values of l is found to be as follows:

l/mm	100	200	300	400	500
f/Hz	720	385	270	205	165

Plot a graph of l against $1/f$ and deduce a value for the speed of sound in the tube. Is the end correction negligible?

24-26 An open ended tube of length 0.30 m is excited in its first three modes of vibration. Calculate their frequencies. (Assume that end-corrections are negligible). If the amplitudes of the overtones are respectively one half and one quarter of that of the fundamental use the principle of superposition to draw a rough y–t graph for the oscillations at a point close to the end of the tube.

24-27 Describe how you would measure the frequency of vibration of a tuning fork. Comment on the uncertainties in your measured quantities.

24-28 In the figure the steel bar of length 0.75 m is clamped at its centre. The glass tube contains an air column the length of which can be altered by an adjustable plunger. The steel bar is stroked along its length with a resined cloth so as to set up a longtitudinal standing wave with a node at its clamped central point and antinodes at its ends. With suitable adjustment of the plunger, it is found that a fine powder placed in the tube forms regularly spaced heaps. The speed

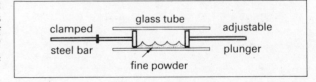

of propagation of longitudinal waves in a long, thin metal bar of Young modulus E and density ρ is given by $c = \surd\,(E/\rho)$.
(a) Find the frequency of the note emitted by the steel bar.
(b) Explain why the powder forms into heaps. Why is it necessary to adjust the position of the plunger before the heaps form?
(c) If the average spacing of the heaps of powder is 49 mm, find the speed of sound in air.
(d) When the experiment is repeated with a clamped brass rod of the same length, the average spacing of the heaps is 74 mm. Find the Young modulus of brass. (The Young modulus of steel $= 2.1 \times 10^{11}$ Pa; ρ for steel $= 7.8 \times 10^3$ kg m^{-3}; ρ for brass $= 8.5 \times 10^3$ kg m^{-3}.)

24-29 Two open-ended organ pipes are sounded together and 4 beats per second are heard. If the longer pipe is of length 0.85 m what is the length of the other pipe? You may ignore any end corrections.

24-30 A descant recorder has a tube which is 225 mm long. The air in it vibrates as in a tube open at both ends. What can you deduce about the range of notes which can be played on the recorder from this information?

Sound and hearing

24-31 A small loudspeaker emits *sound* energy uniformly in all directions in front of it with a total power of 5.0×10^{-5} W. What is the sound intensity (the rate of flow of sound energy per unit area) at a distance of (a) 2.0 m and (b) 12 m from the speaker? How does the amplitude of molecular vibration in the air around the speaker depend upon distance r from the speaker?

24-32 The maximum legal loudness for a car in Great Britain (measured under specified conditions) is 80 dB(A). It is found that a certain car exceeds this limit by 3 dB(A). The owner claims that the excess is negligible—'less than 1 part in 25', while the local police official claims that an increase in loudness from 80 dB(A) to 83 db(A) represents a doubling of sound intensity. Discuss who is right in this case.

25 Wavefronts and rays

Data refractive index of air $= 1.00029$
refractive index of water $= 1.33$
$c = 3.00 \times 10^8$ m s^{-1}

Introduction

25-1 The heavy line in the diagram represents a single wavefront. The fine arc is a secondary wave from P of radius ct. Copy this diagram and draw a series of secondary waves from Q, R, S . . . etc. in order to predict the shape of the wavefront after at time t. The object on theright is at a distance of $\frac{1}{2}\,ct$ from the original wavefront.

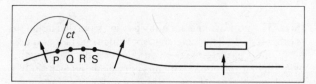

25-2 The Earth's orbit is about 1.49×10^{11} m from the centre of the Sun and the energy flux of the Sun's radiation at this distance is 1360 W m^{-2}. Calculate the solar energy flux at the orbit of the planets (a) Mercury and (b) Jupiter, whose mean distances from the Sun are approximately 0.58×10^{11} m and 77.8×10^{11} m respectively.

25-3 Suppose that the wave energy resulting from a person jumping into a swimming pool is 160 J.
(a) Calculate the energy per unit length of the wavefront when it has a radius of (i) 3.0 m and (ii) 6.0 m.
(b) If the amplitude of the wave is 50 mm at 3.0 m what is it at 6.0 m? Comment on any assumptions you make.

Reflection

25-4 A plane wave moving at 0.40 m s^{-1} approaches a flat reflecting barrier at an angle of incidence of 30°. Draw a series of diagrams at intervals of 0.25 s to show the progress of a piece of the wavefront AB, 1.0 m long, from the moment A reaches the barrier to the moment B reaches it.

25-5 Draw a diagram showing a ray of light being reflected by a plane mirror, the angle of incidence being 60°. Through what angle is the reflected ray deviated? If the mirror is rotated until the angle of incidence is 45°, through what angle is the reflected ray rotated? Now prove the general case, i.e. show that for a mirror rotation of α the reflected ray is rotated 2α.

Refraction

25-6 Plot a graph of the angle of refraction θ_g against the angle of incidence θ_a for values of $\theta_a = 0°, 10°, 20°, 30°, 45°, 60°$ and 90° for a ray of light refracted from air into glass of refractive index 1.50.

25-7 A ray of light travelling in water strikes its horizontal surface at an angle of incidence of 40°. Find the angle θ_a at which it emerges into the air.
  If a layer of oil of refractive index 1.44 is poured onto the water surface the ray emerges into the air at an angle θ. Calculate the path of the ray in the oil and show that $\theta = \theta_a$.

25-8 The speed c of water waves in a ripple tank can be written (approximately) as $c = k\sqrt{h}$ where k is a constant. Plane waves, moving at 0.25 m s^{-1} in a region where $h = 25$ mm, are refracted at a straight boundary at which the depth h changes to 4.0 mm.
(a) Calculate the speed of the refracted waves.
(b) What is the angle of deviation of the waves (the angle between the incident and the refracted wavefronts) if the incident waves meet the boundary between deep and shallow water at an angle of $45°$?

25-9 In an experiment to measure the speed of light a light pulse is reflected up and down a straight tube which is 1200 m long. How much longer will the light take to make 20 journeys there-and-back if the tube is full of air than if it is evacuated? [See data.]

25-10 The refractive index of air varies with temperature. Describe two different everyday observations which support this statement.

25-11 A narrow beam of light enters one long side of a rectangular block of glass at an angle of incident of $45°$. Calculate the angle of refraction in the block if the refractive index of the glass is 1.54. If the block is 80 mm wide find by drawing or by calculation the *sideways* displacement of the beam when it emerges into the air.

25-12 Because of refraction in the Earth's atmosphere, the light from a star which is observed to be $19°\,54'$ above the horizontal enters the Earth's atmosphere at a smaller angle θ as shown in the figure. Calculate θ and thus find the angular error $\Delta\theta$ in the observed altitude (above the horizon) of the star. (See data for n_{air} at ground level.) Is $\Delta\theta$ the same for all observed stars?

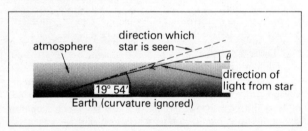

25-13 A glass prism has a refracting angle of $60°$. The graph

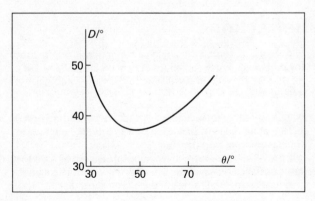

shows how the deviation D varies with the angle of incidence θ of a ray of light entering the prism. (a) What is the minimum deviation D_m? Use the graph to explain why the minimum deviation must occur when the light passes symmetrically through the prism. (b) Calculate the refractive index of the glass of the prism.

25-14 From the graph of D against θ in the previous question find the two angles of incidence for which the deviation is $45°$. Draw rough diagrams to show the paths of rays incident at these two angles through the prism. What do you notice about your diagram?

25-15 A narrow beam of microwaves is travelling in pitch, a material in which their speed is 1.78×10^8 m s^{-1}. They are aranged to be incident on the flat surface of the pitch at various angles of incidence θ_p given by $\theta_p = 20°$, $30°$, $40°$ and $50°$. Draw labelled diagrams showing the beam reflected within the pitch and refracted into the air for each value of θ_p, calculating the relevant angles.

25-16 What is the difference between the critical angles of red light and blue light for a glass for which $n_{blue} = 1.639$ and $n_{red} = 1.621$? How could you demonstrate that the critical angles were different?

25-17 The figure shows a narrow vertical beam of light incident at grazing incidence on the glass side of a beaker of water. Will the beam emerge into the air above the water or be totally reflected at the water-air surface?

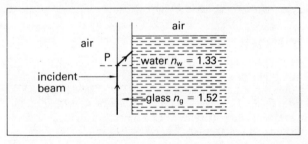

25-18 The critical angle for a ray passing from one liquid to another is $53.1°$. The wavelength of light in one of the liquids is 590 nm. Calculate the possible values for the wavelengths of this light in the other. Explain your calculations carefully.

25-19 Water waves of frequency 8.0 Hz, travelling in a region where the wave speed is 0.32 m s^{-1}, are obliquely incident on the boundary of a region of deeper water where the wave speed is 0.40 m s^{-1}. Sketch a wavefront diagram of the situation if the angle of incidence is equal to the critical angle θ_c. Calculate the value of θ_c in this case.

25-20 A ray is incident perpendicularly on one face of a glass prism of refracting angle A and refractive index 1.50. Calculate the deviation D of the ray for values of A given by $A = 20°$, $15°$, $10°$, $5°$ and $1°$. For these values plot a graph of D against A. What is the slope of the graph for small values of A? Does your answer agree with the relation $D = (n - 1)A$?

25-21 An observer looks at a vertical line through a glass prism of refracting angle 5.0° and refractive index 1.54. The prism is held close to his eye with the refracting edge vertical. How far from the line is he standing if the line appears to be displaced sideways through a distance of 0.10 m?

25-22 An 'air-cell' is made from two microscope slides which are cemented together around their edges to enclose a thin layer of air. The air-cell is placed in a tank of paraffin. A narrow beam of monochromatic light, e.g. from a sodium vapour lamp, is shone through the air-cell and an observer notices that as the cell is rotated the light is cut off when the angle of incidence of the beam on the microscope slide exceeds 42.5°. Draw a ray diagram to illustrate the cut-off situation and calculate the refractive index of the paraffin. [Hint: You will need to introduce the refractive index of the glass of the microscope slides into your calculation but you do not need to know its value.]

Diffraction

25-23 Sketch a series of diagrams to illustrate the behaviour of plane water waves for values of the wavelength λ given by $\lambda/m = 0.1, 0.5, 1.0$ arriving at a barrier in a children's paddling pool in which there is a gap of width 1.0 m. Describe what is happening to the wave energy passing through the gap when $\lambda = 1.0$ m.

25-24 Take two pencils and hold them vertically and *very close* together (but not touching). Look through the narrow gap between them at a vertical edge or bar in a nearby window. Describe what you see. What do you see, again with the gap held vertically, or if you look at a horizontal edge or bar in the window? Explain your observations.

25-25 What do you see when you look at a distant sodium street light, i.e. one several hundred metres away, through the fabric of an open umbrella? In what way does what you see differ when the distant light has a filament bulb?

25-26 (a) A 'harvest moon' is the name sometimes given to the orange appearance of a full moon in late summer when the atmosphere contains a lot of dust. Explain this orange colour.
(b) Why does cigarette smoke have a slightly blueish colour when white light is scattered from it?

Images

25-27 A man 1.60 m tall stands in front of a vertical plane mirror which extends from the floor to a height of 2.00 m. Draw a diagram to show the image of the man in the mirror and calculate the minimum length of the mirror in which he could see all of himself. Assume that his eyes are 1.50 m from the floor. If the top of the mirror is now gradually tilted away from the man explain qualitatively why he would first lose sight of his feet.

25-28 A normal eye cannot focus (for any length of time) a bundle of rays diverging from a point less than 250 mm from it. Draw a ray diagram to show how the eye can best be placed to see the image in a plane mirror of a point object placed 100 mm from the mirror. Add *wavefronts* to the diagram, drawing the arcs carefully with a compass.

25-29 A spark discharge occurs just above ground level 3.0 m in front of a large vertical plane mirror. The flash lasts for 2.0 ns. Draw a sketch showing the regions *on the ground* above which the light energy is moving 20 ns after the start of the flash. Explain how you draw your sketch, marking in any calculated distances.

25-30 Copy figure (a) drawing a semicircle of radius 100 mm. Taking each of your 16 evenly spaced lines to be a ray incident on a concave mirror (the semicircle) draw the normals at the points of incidence and construct a series of reflected rays. They will not form a perfect image of the distant object but will form a cusp shaped figure very much like the bright patch on the surface of a cup of tea cast by sunlight or by the light from a single bulb.

By considering only the central (i) 12 rays (ii) 8 rays and (iii) 4 rays, explain how reducing the aperture of the mirror sharpens the image.

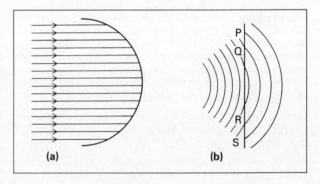

(a) (b)

25-31 Figure (b) shows circular wavefronts which are being refracted at a plane boundary between shallow water and deep water.
(i) Which part of the diagram represents deep water?
(ii) Find the ratio of the wave speeds in deep and shallow water. Explain your calculation.
(iii) Roughly locate the centres of the circles of which the two sets of wavefronts form parts and calculate the ratio of the radii of the two parts of the wavefront PQRS.
(iv) Comment on your answers to (ii) and (iii).

25-32 (a) What is the apparent depth d of water in a swimming pool known to be 3.00 m deep? Assume that the observer is looking straight down into the water.
(b) It is not easy to measure d experimentally. One method is to place a plane mirror on the surface of the water and to look at the image of an object held a distance h above the mirror. When the image exhibits no parallax with a mark on the bottom of the pool the $h = d$. Explain what is meant by no parallax and draw a diagram to describe this experiment.

25-33 The table gives the refractive indices of two glasses for various wavelengths of light in the optical part of the spectrum.

λ/nm	405	486	546	589	656
n_{crown}	1.532	1.523	1.519	1.517	1.515
n_{flint}	1.685	1.664	1.655	1.650	1.644

Plot two graphs with λ along the x-axis to show the variation of n with λ for crown and flint glass. For a thin prism of flint glass with a refracting angle of $6.00°$ calculate (a) the deviation of (i) red light: $\lambda = 650$ nm (ii) blue light: $\lambda = 450$ nm. (b) What is the difference of the deviations, i.e. the angle of dispersion produced by the prism?

25-34 The flint glass prism described in the previous question is used with a crown glass prism of refracting angle A to produce a combination which deviates white light but produces *no net dispersion* between the red and the blue parts of the spectrum.
(a) Draw a sketch to show the arrangement of the prisms with red and blue light passing through them and calculate A.
(b) What deviation of the red and blue light is produced by the combination of prisms?

26 Optical systems

Data $c = 3.00 \times 10^8$ m s^{-1}

least distance of distinct vision for a normal eye $D = 250$ mm

Lenses

26-1 A converging lens has a forcal length of 0.20 m. Find the nature (real—R or virtual—V), position and size of the image of an object 5.0 mm high which is placed (a) 1.00 m (b) 0.40 m (c) 0.22 m (d) 0.15 m from the lens. State whether the image is upright (U) or inverted (I) in each case.

26-2 Repeat each part of the previous question for the case when the converging lens is replaced by a diverging lens of focal length 0.20 m.

26-3 A naturalist wishes to photograph a rhinoceros which is 75 m away. The beast is 3.8 m long, and its image is to be 12 mm long on the film. (a) What focal length lens should he use? (b) What would the image size have been if he had used a normal 50 mm focal length lens?

26-4 A converging lens of focal length 0.18 m is used to produce an image of an illuminated object on a screen. If a magnification of 9.0 is required, determine (a) the distance from the object at which the screen must be placed (b) the position of the lens in relation to the object and the screen.

36-5 A man wishes to observe fine detail on a photograph using a hand-held lens as a simple magnifying glass in such a way that he sees a ten-times magnified image 250 mm from the lens. What focal length lens should he use and how far from the photograph should it be held?

26-6 The diagram shows a plane wavefront at near normal incidence on a thin prism of refracting angle A made of glass of refractive index n. By considering the distance nl travelled in air by the part of the wave incident at the apex of the prism compared to the distance l travelled in the prism by the wave incident a distance b below the apex, show that the deviation of the wavefront is given by $D = (n - 1) A$.

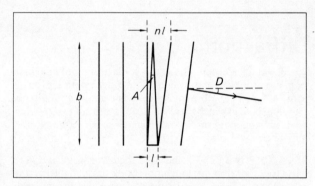

26-7 The converging lens in a typical miniature camera has a focal length of 50 mm. The lens can be moved relative to the film in order to focus objects from inifinity down to about 0.50 m. What is the range of movement of the lens?

26-8 Tabulate a series of values of u and v for a converging lens for real objects and images. Sketch graphs of (a) v/u against v (b) $1/v$ against $1/u$ (c) $u + v$ against u, for a converging lens of focal length 0.20 m. [Plot the first mentioned quantity on the y-axis in each case.]

26-9 A plane mirror M, a converging lens L and a source of light consisting of an illuminated metal gauze S mounted across a hole in a card are placed alongside a metre rule attached to the bench. With M at 822 mm, L at 785 mm and S at 582 mm a clearly focused image of S appears on the card. The image is again clearly focused when M is moved first to 863 mm and then to 885 mm. Draw ray diagrams to explain how the image is formed and calculate the focal length of L.
  What contributes to the uncertainty in the measured value of f? How big might the uncertainty be in this experiment?

26-10 A telephoto lens system consists of a converging lens of focal length 115 mm placed 100 mm in front of a diverging lens of focal length 30 mm. Find the position of the image formed of a distant object. Draw a diagram showing rays from the distant object (which form a beam parallel to the axis by the time they reach the first lens) passing through the lens system.

26-11 A thin converging lens of focal length 0.30 m is placed in contact with another converging lens of focal length 0.20 m. Show that the image of a distant object is 0.12 m from the lenses, i.e. that the lens combination has an effective focal length of 0.12 m.

26-12 A small bulb and a screen are placed 1.40 m apart. A converging lens moved between them is found to produce a real image of the bulb on the screen at each of two lens positions L_1 and L_2. If $L_1L_2 = 0.48$ m, calculate the focal length of the lens. [Hint: let the distance from the bulb to L_1 be x.]

26-13 Calculate the position and magnification of the image formed by two coaxial converging lenses, each of focal length 0.100 m and placed 0.300 m apart, of an object placed on the axis 0.125 m from one of the lenses. Draw a sketch showing the formation of the image to support your calculation.

Curved mirrors

26-14 A concave mirror has a radius of curvature of 0.30 m. Find the nature (real or virtual), position and size of the image of an object 10 mm high which is placed (a) 10 m (b) 0.30 m (c) 0.10 m from the mirror. State whether the image is upright or inverted in each case.

What would be the result in case (b) if the mirror was broken and only about one third of its original reflecting surface was used to form an image?

26-15 Repeat each part of the previous question if the concave mirror is replaced by a convex mirror of radius of curvature 0.30 m.

26-16 You are provided with a diverging lens of focal length f (where f is believed to be between 0.1 m and 0.2 m), a concave mirror of radius of curvature 0.10 m, a small bright source and a screen. Explain how you could use the apparatus to measure f.

26-17 A small, bright object is placed 0.22 m from a converging lens of focal length 0.15 m. A convex mirror of radius of curvature 0.32 m is placed on the other side of the lens from the object and coaxial with it. Calculate the distance between lens and mirror when a real inverted image is produced alongside the object.

26-18 A plano-convex wax lens is found to have a focal length of 0.44 m for microwaves of wavelength 38 mm. If the radius of the curved surface of the lens is 0.32 m calculate the refractive index of the wax for microwaves.

26-19 A compound lens is to be made from an equi-convex crown glass lens and a plano-convex flint glass lens, the curved surfaces of which all have the same radius r. If the refractive indices of flint and crown glass are 1.65 and 1.52 derive a relationship between the focal length f of the compound lens and r. Hence show that for a focal length of 1.00 m, r should be 0.39 m.

26-20 A set of plane wavefronts, travelling in air with their planes perpendicular to the axis of the wax lens described in question **26-18**, strike its plane face. If the lens is 66 mm thick at its centre draw a diagram roughly to scale to show how these wavefronts are refracted through the lens.

26-21 For an equiconvex thin lens, of which the radii of curvature of the faces are each 462 mm, the refractive indices for red light and for blue light are: $n_{red} = 1.64$ and $n_{blue} = 1.68$. A small white light source is placed on the axis of the lens and 500 mm from it. Calculate the positions of the images formed by (a) the red light (b) the blue light from the source.

26-22 The figure shows the image formation described in the previous question. The dashed lines represent the rays of red light and the full rays the blue light. Describe as fully as possible what is seen on a screen placed (a) at S_b and (b) at S_r if the lens has an aperature of 60 mm. (You will need to use the answers to the previous question.)

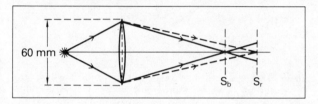

The eye

26-23 A man with eyeballs of effective length 25 mm, i.e. with this distance from refracting region to retina, walks towards a white post which is 4.0 m high. Calculate the size of the image of the post on his retina when he is a distance d from the post given by $d/m = 400, 100, 20, 5$.

26-24 The Moon subtends an angle of 0.52° at the Earth's surface. What is the size of the image of the Moon on the retina of an eye of length 21 mm from refracting system (the cornea plus lens) to retina? Describe how (a) the size and (b) the intensity of the image varies when the pupil of the eye changes from 2.0 mm to 6.0 mm in diameter.

26-25 The power of a normal eye is $\approx +40$ D when it is relaxed and it can accommodate by about +4 D. Amplify and explain this statement as fully as possible.

26-26 What is the visual angle subtended by the sides of this letter

H

when the pages is 0.75 m from your eye? How much bigger would the visual angle be if the page was held at the least distance of distinct vision for a normal eye?

Estimate the greatest distance at which you can 'read' the H as that letter and hence the visual angle at which your eyes can just resolve the sides of the H.

26-27 A copper sulphate crystal measuring 2.0 mm by 1.0 mm is placed at the principal focus of a converging lens of focal length 80 mm. Calculate the angles subtended by the sides of the image at the eye and show that these are just over three times the visual angle subtended by the sides of the crystal placed at the near point of an unaided eye.

26-28 In the figure an object O of height 40 mm has an image I at the near point of a normal eye E. If the lens is of focal length 0.20 m and is 0.55 m from the eye find the linear magnification m of the image. Calculate (a) the visual angle α_i of the image and (b) the visual angle α_o of the object as seen by the eye with the lens removed. Hence find the angular magnification M of the system.

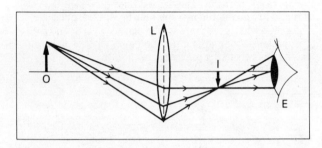

26-29 The eye can just resolve two objects which are separated by an angle of about 3×10^{-4} rad. A magnifying glass has a maximum useful magnifying power of about 10. Can a single lens be used to count the scratches on a diffraction grating with 100 lines per millimetre? [Calculate the visual angle subtended at an unaided eye by two adjacent scratches on the grating with the grating at the near point.]

26-30 With the help of a simple ray diagram explain the formation of a real image of a very distant object (e.g. the Moon) by a converging lens. What factors determine (a) the diameter and (b) the brightness of this image? If the lens has a power of +2.5 D and the image of the Moon is 3.6 mm in diameter, calculate the angle subtended by the Mon at the Earth's surface.

26-31 A thin converging lens of focal length 50 mm is laid on a map situated 605 mm from the eyes of an observer. Describe what is seen when the lens is raised (a) 50 mm (b) 55 mm (c) 60 mm above the map. Where possible calculate the angular magnification, i.e. the ratio (angle subtended by two points on the map viewed through the lens)/(angle subtended by the same points when the map is viewed at the near point with the unaided eye).

Optical instruments

26-32 A two-lens microscope consists of two converging lenses: an objective lens O of focal length 40 mm and an eyepiece lens E of focal length 100 mm. If a small object is placed 50 mm in front of O on the common axis of the lenses find
(a) the size and position of the image formed by O

(b) the distance of E from O if the final image is to be virtual and 250 mm from E
(c) the overall linear magnification of the microscope.

26-33 An object 1.6 mm high is placed as shown in the figure between two converging lenses O and E. If $f_o = 20$ mm and $f_e = 50$ mm calculate the position, nature and size of the images of the objects formed by the two lenses.

The pair of lenses is used as a microscope with the observer's eye to the right of E. Show that he sees a 25-times magnified image, 200 mm to the left of E, when an object is placed 23 mm to the left of O.

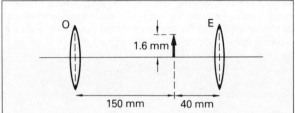

26-34 Why can a microscope resolve detail better when the object being studied is illuminated with blue light rather than under ordinary conditions of illumination?

26-35 An astronomical telescope has an objective lens with a focal length of 1.45 m and an aperture of diameter 80 mm. Its eyepiece lens has a focal length of 0.12 m. With the incident light parallel to the axis and the final image at infinity, calculate the minimum diameter of the aperture of the eyepiece lens if all of the light from the objective lens is to pass through the eyepiece lens and reach the observer.

26-36 The focal length of the objective lens of the Yerkes Observatory's largest refracting telescope (the biggest of its kind in the world) is 20 m. A photograph of the Moon is made using this telescope. How should the photographic plate be mounted and how large must the plate be in order to record the whole of the Moon's image? The moon is 3.5×10^3 km in diameter and is 3.8×10^5 km from the Earth.

26-37 If the Yerkes Observatory telescope described in the previous question uses an eyepiece lens system of effective focal length 36 mm calculate the angular magnification of the telescope with the final image at infinity.

26-38 Large telescopes use concave mirrors as their objectives. What would be the effect on the real image formed by such an objective of separately (a) doubling its aperture, e.g. from 250 mm to 500 mm (b) doubling its focal length, e.g. from 3.0 m to 6.0 m?

26-39 A camera with a lens of focal length 60 mm is used to take a colour slide of a distant scene. The slide is held up to the light and viewed with the unaided eye from a distance of 200 mm. Consider a tree which subtends an angle of 0.13 rad at the photographer. Calculate (a) the size of the tree's image on the slide (b) the angle subtended at the eye of the person who views the slide (c) the overall magnifying power is obtained. Draw diagrams to explain your calculations.

26-40 The lens of a relaxed eye has an effective focal length of 21 mm. The diameter of the eye pupil varies from 1.5 mm to 6.5 mm. What are the corresponding f-numbers of the eye? How many times greater is the intensity of light incident on the eye when the pupil contracts to 1.5 mm than when it dilates to 6.5 mm? (Assume the pupil adjusts to give an image on the retina of constant brightness.)

26-41 Discuss the problems involved in designing and producing a large-aperture refracting telescope. To what extent are these problems overcome or reduced by the use of a concave reflecting mirror in place of a converging objective lens?

26-42 A converging lens of focal length 1.20 m forms an image of a distant tree as shown in the figure. Calculate (a) the angle subtended by the tree at the lens and, if the tree is 400 m from the lens, (b) the height of the tree. This image is arranged on separate occasions to lie at the principal focus of P, a converging lens and Q, a diverging lens, each of focal length 0.20 m placed as shown. Calculate (c) the visual angle of the final image viewed through each of these lenses in turn. What is (d) the angular magnification of the telescope formed in each case? (The arrangement with lens Q is called a Galilean telescope.)

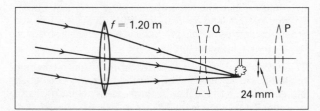

26-43 A laboratory telescope is made from two converging lenses: an objective O of focal length $f_o = 0.80$ m and 60 mm in diameter, an eyepiece E of focal length $f_e = 0.10$ m and 40 mm in diameter. O and E are placed 0.90 m apart. Calculate the size and position of the image I of lens O formed by lens E. (This is sometimes called the *exit pupil* of the telescope and all light incident on O passes through I.) Does the size of I depend on the diameter of lens E? Will an eye pupil placed at I catch all the light which enters the telescope through O?

The spectrometer

26-44 A prism of refracting angle A is placed on the table of a spectrometer and light from the properly adjusted collimator is viewed when reflected in turn from each of the sides of the prism forming A. The readings on the telescope vernier are $146°\,44'$ and $26°\,48'$. Calculate the angle A.

26-45 A spectrometer is set up and properly adjusted for use with a prism. Describe how you would measure the angle of minimum deviation produced by a prism for sodium yellow light.

26-46 A prism of refracting angle $59.88°$ is placed on a spectrometer table and the angle of minimum deviation for yellow (sodium D) light is found to be $44.52°$. Calculate the refractive index of the glass of the prism for yellow light to four significant figures.

27 Patterns of superposition

Data $c = 3.00 \times 10^8$ m s^{-1}
speed of sound in air $= 340$ m s^{-1}

Coherence

27-1 Some lasers produce a pulsed beam of light. If the average power of such a beam is 60 mW and it consists of 200 pulses per second each lasting for only 1.0 μs, what is the energy in each pulse and the power of the laser beam during the pulse?

27-2 A point source of microwaves of wavelength 32 mm is placed 600 mm from the first of two parallel sheets of cardboard which are separated by 72 mm. Express the distance between the images of the source produced by the partial reflections at the two surfaces of the film (a) in millimeters and (b) in terms of the wavelength λ of the microwaves. Repeat (a) and (b) for the case where the source is 300 mm from the first sheet of cardboard.

27-3 Two in-phase coherent sources of sinusoidal waves are placed at the points S_1 and S_2 in the figure. A detector registers a maximum at both P and Q and when moved from P to Q detects a single minimum between them. Deduce the

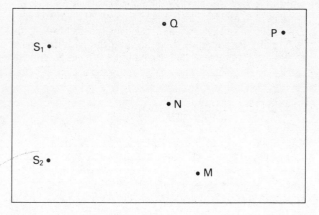

wavelength of the waves emitted by S_1 and S_2. [You will need a ruler. Assume the figure is reduced by a factor of ten from the real situation.] Describe and explain what the detector will register when placed at (a) N and (b) M.

27-4 The photograph shows a typical ripple tank two-source superposition pattern. If the scale is such that 10 mm on the photograph is equivalent to 120 mm on the water surface, calculate the wavelength of the water ripples. By measuring distances such as S_1P and S_2P, verify that the bright patches on the photograph represent places where the two wavefronts from S_1 and S_2 arrive in phase.

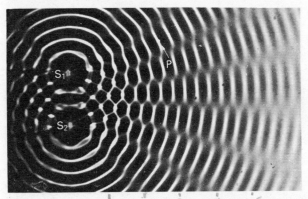

27-5 Two ripple tank dippers S_1 and S_2 are vibrating *in antiphase*. Draw (you will need a compass) a series of arcs to represent the wavefronts from S_1 and S_2, using a wavelength λ such that $S_1S_2 = 3.5\ \lambda$. Mark on your diagram the nodal and antinodal lines and state how this pattern of superposition differs from that where $S_1S_2 = 3.5\ \lambda$ but S_1 and S_2 are in phase.

Young's experiment

27-6 In a Young's slit type of experiment with microwaves of wavelength 30 mm the distance between maxima detected at a distance of 42 m from the sources while moving across the line of symmetry was found to be 2.5 m. What was the separation of the two microwave sources?

27-7 This is a question about the geometry of the two-source Young's fringes experiment. [You will need a calculator with an 8-digit display.] In the figure which is *not* to scale sources at S_1 and S_2 send light to a point P. If $a = 0.400$ mm, $D = 0.800$ m and $x = 3.00$ mm, calculate the lengths S_2P and S_1P (using Pythagoras's theorem) and the path difference $(S_2P - S_1P)$ between light arriving at P from S_1 and S_2.

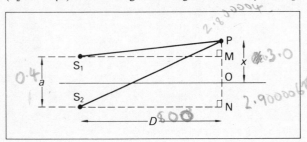

Explain whether for light waves of wavelength 600 nm from coherent sources at S_1 and S_2 there will be a maximum or a minimum intensity at P.

27-8 Repeat the previous question algebraically, i.e. express S_1P^2 and S_2P^2 in terms of a, D and x and evaluate
$$S_2P^2 - S_2P^2$$
By writing this as $(S_2P - S_1P)(S_2P + S_1P)$ and approximating the latter term to be $2D$, prove that
$$\frac{ax}{D} = n\lambda \quad \text{(where n is an integer 1, 2, 3 ... etc.)}$$
for there to be a maximum of intensity at P for light of wavelength λ.

27-9 Using the photograph which accompanies **27-4** calculate the wavelength λ of the ripples by measuring the distance between the sources a, and the ratio $\Delta x/D$ where Δx is the separation of the maxima at a distance D from the sources. Take the scale of the photograph to be 1/12 real size and use the relationship $\lambda \approx a\Delta x/D$. In the ripple tank experiment λ was 48 mm. Does your calculation of λ support the use of the (approximate) formula in this case?

27-10 An experimenter wishes to demonstate the wave nature of the output from a small ultrasonic generator consisting of a quartz crystal oscillating at 50 kHz. Explain how he could arrange the apparatus to get a two-source interference pattern and give approximate dimensions of the arrangement which would produce intensity minima separated by 200 mm at his detector.

27-11 Two loudspeakers L_1 and L_2 driven from a common oscillator are set up as shown in the figure. A detector is placed at D. It is found that, as the *frequency* of the oscillator is gradually changed from 200 Hz to 1000 Hz the detected signal passes through a series of maxima and minima. Explain why this is so. Calculate the frequency at which the first minimum is observed.

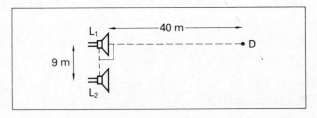

27-12 Microwaves, when they are reflected from a metal sheet, and light, when it is reflected at the front surface of a piece of glass, both undergo a phase reversal, i.e. a phase change of π rad. Outline one experiment for each of these types of electromagnetic radiation which would provide support for this statement.

27-13 The only practical way to produce visible patterns of superposition (interference patterns) with light is, in effect, to derive two sources from a single source. Explain why this is so and describe how diffraction may be used to produce such a double source.

27-14 The figure shows a transmitter of electromagnetic waves of wavelength $\lambda = 0.30$ m placed at T a distance of 6.0 m from a receiver at R. A plane reflecting surface M is held as shown in such a way that the perpendicular distance from M to the direct line TR is 2.0 m.

Express the path lengths of the two wavetrains from T to R in terms of λ. If the wave which reflects from M undergoes a phase change of π rad are the two waves arriving at R in phase or in antiphase?

Describe as fully as possible how the signal received at R varies as M is slowly moved towards the line TR until it almost lies along it.

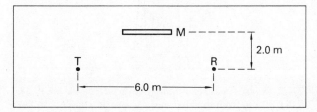

27-15 The speed of light in air is 0.999 75 times its speed in a vacuum, which is $2.997\,9 \times 10^8$ m s^{-1}. Use this information to calculate the number of oscillations of yellow light of frequency $5.085\,3 \times 10^{14}$ Hz which occur when the light travels down a tube which is exactly 30 mm long (a) when the tube contains air and (b) when the air is evacuated from it.

If the air is gradually evacuated the phase of the emerging light will slowly vary. Explain this statement and calculate the change of phase between conditions (a) and (b) above.

27-16 In the figure P and Q are two radio navigation stations transmitting continuous sinsusoidal radio waves at a wavelength of 3500 m. The stations are 70 km apart and transmit signals which are in phase and of equal amplitude. A ship moves from Y, equidistant from P and Q, towards P.
(a) Describe how the received radio signal varies during the first 3 km of its journey.
(b) Another ship is at X where it receives a maximum signal. It moves on a line XZ which is parallel to YQ. How does the received signal during the first 3 km of its journey compare with that of the ship moving from P?

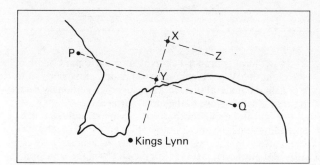

Thin film patterns

27-17 The figure shows two microscope slides arranged so as to enclose a wedge-shaped air film. The slides are illuminated with monochromatic light and an interference pattern consisting of a series of equally spaced fringes parallel to the line where the slides meet is observed. Explain what would be seen if, separately, the following changes were made:
(a) the metal foil was moved very slowly to the right
(b) the top microscope slide was exchanged for one which was not quite flat
(c) the space between the microscope slides was filled with water
(d) the wavelength of the light used to illuminate the microscope slides was doubled
(e) the temperature of the metal foil was raised.

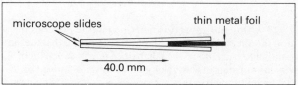

microscope slides   thin metal foil

40.0 mm

27-18 If in the previous question the distance occupied by 20 fringe-spacings was measured and found to be 4.9 mm, and the wavelength of the light used was 590 nm, calculate the thickness of the metal foil. (Assume that the illumination and the viewing are from directly above the slides.)

27-19 A wedge-shaped film of air between two flat glass plates gives equally spaced dark fringes, using reflected sodium light, which are 0.12 mm apart. When monochromatic light of another wavelength is used the fringes are 0.14 mm apart. Explain why the two fringes separations are different and calculate the wavelength of the second source of light. (Assume that the incident light falls normally on the air film in both cases and take the wavelength of sodium yellow light in air = 590 nm.)

27-20 Describe an experimental arrangement to observe the pattern of superposition produced when the air film between a long focal length converging lens and a flat glass block is illuminated from above with monochromatic light. If the wavelength of the light used is 586 nm explain what is happening at a place where the air gap is 4.7 μm thick.

27-21 If in the previous question there is perfect contact between the lens and the block at the centre of the interference pattern, the radius r_n of the nth bright circle is given by the equation

$$r_n^2/R = (n + \tfrac{1}{2})\,\lambda$$

where R is the radius of curvature of the lens surface and λ the wavelength of the light used. In an experiment in which R was 1.50 m the following values of n and r_n were measured:

n	2	5	10	15	20
r_n/mm	1.45	2.14	2.98	3.60	4.16

Plot a graph of r_n^2 against n and use your graph to find λ.

27-22 A microwave transmitter T and receiver R are placed side by side as shown in the figure facing two sheets of material N and M. It is found that a very small signal is registered by R; what can you deduce about the experimental set up?

When M is moved towards N a series of maxima and minima is registered by R. Explain this and deduce the wavelength of the electromagnetic waves transmitted by T if the distance from the 2nd to the 7th minima is 70 mm.

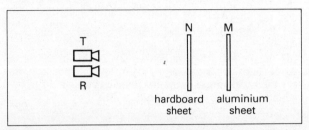

27-23 Using the apparatus described in the previous question the sheets of material M and N are fixed so that they are parallel to each other with a gap of 22 mm between them. T, which transmits at a wavelength of 28 mm, is moved so that the beam of microwaves it produces strikes N obliquely at a glancing angle θ. R is swung round so as to receive waves reflected from N and M. Sketch this new arrangement and find the angle at which a maximum signal will be received by R.

Diffraction at a slit

27-24 A plane sound wave of frequency 8.5 kHz is incident on a large heavy board in which there is a slit of width 0.10 m. Sound is diffracted through the slit and a small omnidirectional microphone is moved about in the region beyond the slit. Draw a sketch to show the direction(s) in which the microphone detected a minimum of sound.

27-25 The figure shows the diffraction pattern produced by a single slit illuminated by a parallel beam of light of wavelength λ. If the width of the diffracting slit is 0.20 mm, calculate λ.

A second slit, also of width 0.20 mm, is placed alongside the first slit so that the centre-to-centre separation is 0.60 mm. Sketch a graph showing the new relative intensity pattern produced by the slits.

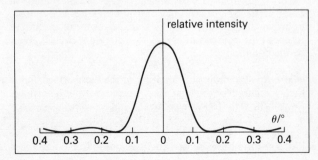

27-26 The figure shows a graph of intensity against distance on a screen placed 2.0 m away from an illuminated double slit in a typical 'Young's slits' experiment. Explain the general shape of the curve (the full line).
(a) If the light was of wavelength 590 nm and the minima on the full curve occur every 4.2 mm, what was the slit separation?
(b) Calculate also the width of *each* slit.
Explain any approximations in your calculation.

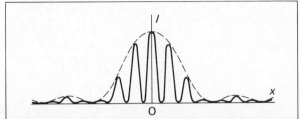

27-27 A laser beam is used to illuminate a flag at night. The flag is 400 m from the laser and is approximately 2 m square. The laser produces light of wavelength 6.6×10^{-7} m and the initial beam width is 2.0 mm. Will the laser beam adequately illuminate the flag? [Assume that the light spreads out by diffraction through the circular end of the laser.]

27-28 Explain briefly what each of the words *diffract, superpose* and *resolve* means in the study of light.

27-29 An optical telescope with an aperture of 250 mm is used to observe two stars which are known to have an angular separation of 3 μrad. Can they be resolved? Take an 'average' value of 600 nm for the wavelength of the light of the stars.

27-30 A parabolic reflecting dish is used to reflect a radar beam emitted from a source placed at its focus. The figure shows how the intensity in the beam varies with direction.
(a) Explain how this figure relates to the graph which accompanies question **27-25**.
(b) If the dish has a width of 1.05 m and the radar system operates at 10 GHz calculate the angular width 2θ of the main beam.

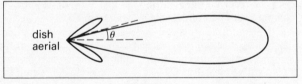

27-31 A weak parallel beam of monochromatic light of wavelength 580 nm is arranged to fall perpendicularly onto a polished flat metal surface. Explain how a stationary wave pattern is set up close to the metal surface.

The antinodes in the pattern can be demonstrated by spreading a thin layer of photographic emulsion of refractive index 1.38 on the metal. Calculate the distance between adjacent antinodal layers (a) in the air (b) in the emulsion.

The diffraction grating

27-32 A diffraction grating is set up on a spectrometer table so that parallel light is perpendicularly incident on it. For light of wavelength 589 nm the first order spectral lines are observed in directions making angle $\theta = 22.0°$ with the straight through position.
(a) Calculate the value of the slit separation s in the grating and the number of slits per millimetre.
(b) Calculate the wavelength of light which would give a first order spectral line at $\theta = 24.3°$.

27-33 How many spectral orders are produced within 15° of the incident direction when a parallel beam of yellow light of wavelength 590 nm falls on a coarse diffraction grating the lines of which are 18 μm apart?

27-34 Plane water ripples of wavelength 5 mm arrive at a barrier which lies parallel to the waves and has small gaps in it at intervals of 12 mm. Show, by drawing a *scale diagram*, that there will be first and second order diffracted waves but no third order.

27-35 'When white light passes through a diffraction grating the grating *disperses* the light. Each slit in the grating *diffracts* the light and the resulting superposition of light from different slits leads to a pattern of peaks or maxima of light intensity at angles of *deviation* which vary from colour to colour.' Write brief notes, including diagrams where appropriate, to explain in your own words each of the italicised words.

27-36 A diffraction grating having 5.00×10^5 lines m^{-1} is illuminated with a parallel beam of white light of extreme wavelengths $\lambda_r = 750$ nm and $\lambda_v = 400$ nm.
(a) Calculate the angular dispersion between the red and violet in the first order spectrum.
(b) A lens of focal length 200 mm is used to form an image of this first order spectrum in order to record it photographically. What will be the distance between the red and the violet ends of the spectrum on the film? Support your calculation with a diagram.

27-37 Show that the second and third order spectra of visible light produced by a diffraction grating will *always* overlap regardless of the number of lines per millimetre in the particular grating used. (Take values of λ_r and λ_v from the previous question).

27-38 A sodium lamp is viewed through a diffraction grating held close to the eye. The lamp is 2.5 m from the grating. Two images of the lamp are seen, one on either side of the position of the lamp and each about 0.6 m from it as read from a tape measure laid on the bench directly beneath the lamp. If the grating has 400 lines per millimetre, calculate an approximate value for the wavelength of sodium light.

27-39 Explain how you would use a transmission grating, suitably mounted on a spectrometer, to determine the wavelength of a monochromatic beam of light. (Assume that the collimator and telescope of the spectrometer are properly adjusted.)

27-40 A small detector of infra-red radiation which does not respond noticeably to visible light is moved across a line spectrum produced by a diffraction grating. It detects only one strong signal at a place occupied by a green line of wavelength 520 nm in the second order visible spectrum. What is the wavelength of this infra-red radiation?

27-41 Suggest how you might construct a diffraction grating for use with (a) microwaves of frequency about 10^{10} Hz (b) sound waves of frequency about 10^4 Hz.

27-42 Draw a graph of the intensity against angle of deviation for a two slit superposition pattern using monochromatic light. [See, for example, the graph of question **27-26**.] Below it draw the pattern of intensity against angle of deviation which you would get if (a) one of the two slits were covered up (b) several hundred parallel slits of the same width were added alongside the two slits, with the distance between any pair of adjacent slits equal to the original slit separation.

Crystal diffraction

27-43 The spacing between the layers of atoms parallel to the cleavage face of calcite is 303 pm. A monochromatic beam of X-rays is found to be strongly reflected from a crystal of calcite when the glancing angle it makes with the cleavage face is 14°42′; strong reflection does not occur for any smaller angle than this. Calculate the wavelength of the X-rays. At what other values of the glancing angle might strong reflection occur?

27-44 A crystal is formed by identical atoms packed together in a simple cubic lattice. Draw a two-dimensional sketch of part of the crystal and draw lines to indicate two sets of layers of atoms with different spacings.

A simple crystal in the shape of a cube is placed so that one of its faces is perpendicular to a beam of monochromatic X-rays of wavelength 154 pm. A strong emergent beam is detected at right angles to the incident beam. Explain which layers of atoms are diffracting the X-rays and calculate their lattice spacing.

28 The electromagnetic spectrum

Data $c = 3.00 \times 10^8$ m s^{-1}
$e = 1.60 \times 10^{-19}$ C
the Planck constant $h = 6.63 \times 10^{-34}$ J s
the Stefan-Boltzmann constant
$\sigma = 5.67 \times 10^{-8}$ W m^{-2} K^{-4}

Electromagnetic waves

28-1 Light and X-rays are both said to be electromagnetic waves. What kind of experimental observations suggest that both are wave motions?

28-2 Estimate the time for light to travel (a) the width of an atom (b) across a room (c) from the Earth to the Moon.

28-3 The following data represents some frequencies and wavelengths used in radio and television broadcasting: 200 kHz/1500 m (a.m. radio), 92.6 MHz/3.24 m (f.m. radio), 0.516 GHz/0.581 m (u.h.f. T.V.). Show that the waves associated with these broadcasts all travel with the same speed.

28-4 An electromagnetic radiation of wavelength 0.21 m from hydrogen is of great importance in radioastronomy. What is the energy of a photon of this radiation?

28-5 Mark a logarithmic scale of wavelengths λ down the left side of a piece of paper where λ/m $= 10^4, 10^3 \ldots, 10^{-19}, 10^{-20}$. Mark parallel scales for (a) the frequencies and (b) the photon energies for electro-magnetic waves. Indicate the main regions of the electromagnetic spectrum to the right of your scale showing the range of wavelength, etc. for each region.

28-6 Make a list of the sorts of accelerating process that occur in the production of various types of electromagnetic radiation.

28-7 Outline the methods by which electromagnetic waves of wavelength of the order of (a) 10^{-10} m (b) 5×10^{-7} m (c) 10^{-1} m are detected.

28-8 A plane mirror rotating at 1800 revolutions per minute reflects light on to a stationary mirror 170 m away. The light returns to the rotating mirror, and after the second reflection, is found to follow a path which makes an angle of 4.3×10^{-4} rad with the path it followed when both mirrors were stationary. Calculate a value for the speed of light.

28-9 A disc has 340 small holes cut in a circle near its rim. A narrow laser beam is shone through one of the holes, travels to a mirror placed 6.4 km away and returns through one of the other holes to a photosensitive detecting system which operates a microammeter. The wheel is then spun and the meter reading starts to fall as the speed of rotation is increased. At what rotational speed will the meter next record a maximum value?

28-10 In early measurements of the surface-to-surface Earth–Moon distance d using the reflection of radar waves,

there was an uncertainty of about ± 750 m in d. Assuming that this uncertainty was wholly the result of the measurement of the time taken for the round trip, what was the uncertainty in the measured time?

28-11 A resonance experiment with microwaves of frequency 9.50 GHz gave the distance between adjacent nodes of electric field intensity as 15.7 mm. Calculate the value of the speed of the microwaves which this experiment gives.

28-12 The spectrum of atomic hydrogen contains a bright line at a wavelength of 486 nm. When the spectrum of light from a distant galaxy was analysed this same hydrogen line was measured to have a wavelength of 497 nm. Calculate the speed of recession of the galaxy from the Earth.

28-13 A police speed-trap uses radar waves of wavelength 100 mm. A car moving at 13.4 m s^{-1} (30 m.p.h.) reflects these waves and a change in frequency of f is observed. Calculate Δf. Did you need to know the speed of the radar waves in your calculation?

The nature of e–m waves

28-14 The figure represents a plane electromagnetic wave moving to the right. Describe the figure in detail and thus explain in what way it represents such a wave.

If the maximum value of E is 1.0×10^{-7} V m^{-1}, calculate the maximum electric force experienced by an electron in the path of the wave.

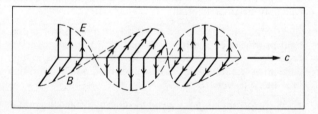

28-15 Draw a graph to show the variation of the E-field with distance along a plane polarised sinusoidal electromagnetic wave of wavelength 1.2 m and amplitude 4.0 µV m^{-1}.
(a) Calculate (i) the frequency of the wave and (ii) the r.m.s. value of its E-field.
(b) What is the r.m.s. potential difference produced by this wave across an aerial 0.60 m long?

28-16 A sodium atom emits a wavetrain of light of wavelength 590 nm. If the time taken by the atom to undergo the energy transition which produced the wavetrain was 3.0×10^{-9} s, how many oscillations of the E-field are there in the photons from sodium atoms?

28-17 A 'ghost' image is seen on a T.V. screen 40 mm to the right of the proper picture. What does this tell you about the received signal if the electron beam covers the entire screen of 625 lines twenty-five times per second and the screen is 0.50 m wide?

28-18 Two radio transmitters T_1 and T_2 emit vertically polarised electromagnetic waves at a frequency of 1.44 MHz. A stationary wave pattern is set up along the line T_1T_2; what is the distance between adjacent nodes in this pattern?

A car is driving along the line between the transmitters (e.g. along the M4) and its a.m. radio is receiving the programme carried by the 1.44 MHz waves. Describe what the driver hears if the car is travelling at just over 30 m s^{-1}.

28-19 Draw a block diagram to show the essential components of a simple radio receiver. Add an energy flow diagram to describe the energy changes which occur starting with an incoming radio wave and finishing with outgoing sound. Indicate any sources of energy and any places at which energy is 'wasted'.

28-20 Explain what is meant by the statement: 'the amplitude of the 30 mm microwaves is modulated sinusoidally at 1000 Hz'.

28-21 The average rate per unit area at which a plane electromagnetic wave transmits energy can be expressed as $E_0B_0/2\mu_0$, where E_0 and B_0 and B_0 are the amplitudes of the electric and magnetic field variations in the wave. Show that the units of this expression are W m^{-2}. ($\mu_0 = 4\pi \times 10^{-7}$ N A^{-2}.)

If the values of E_0 and B_0 at a distance of 10 m from a point source of light are 6 V m^{-1} and 2×10^{-8} T respectively, calculate the power output of the light.

28-22 A steerable dish 77 m in diameter is used as a radio-telescope to study electromagnetic waves reaching the Earth's surface from space.
(a) If the amplifying equipment needs an input power of at least 10^{-9} W, estimate the intensity of the weakest incoming wave which the telescope can detect.
(b) When the telescope is being used to investigate the distribution of atomic hydrogen in our galaxy (the Milky Way), estimate the smallest angle between two areas of hydrogen which it could resolve. (Atomic hydrogen radiates strongly at a wavelength of 0.21 m.)

28-23 Solar radiation arrives at the Earth's orbit at the rate of 1.4 kW m^{-2}. If the average radius of this orbit is 1.49×10^{11} m claculate the power output of the Sun. What rate of mass loss does this represent?

Plane polarised waves

28-24 If you were given a single unmarked sheet of polaroid suggest two methods by which you could establish the direction it should be held to transmit light which is polarised in a vertical plane.

28-25 You are given a box T which you are told is a source of plane polarised electromagnetic waves of about 80 mm wavelength and a second box R which can detect such waves. Describe what simple experiments you would perform with the resources of a school laboratory to test that T is what it is said to be.

28-26 A beam of monochromatic light meets the boundary of a medium of refractive index n at an angle of incidence θ_a; some light is refracted and some is reflected. Show that when the reflected beam is perpendicular to the refracted beam, $\tan \theta_a = n$.

At what angle of incidence will light of wavelength $\lambda = 600$ nm be plane polarised on reflection from flint glass? (The refractive index of flint glass at 600 nm is 1.61.)

28-27 Mica crystals are said to be doubly refracting, i.e. the speed of electromagnetic waves in mica depends on their plane of polarisation. The refractive indices of mica for two mutually perpendicular plane polarised waves of wavelength 589.0 nm in air are 1.605 and 1.612. If unpolarised light passes through a sheet of mica 40.00 μm thick, calculate the number of oscillations undergone by each wave in the mica. What will be the phase difference between the two waves emerging from the mica and what therefore will be the result of their superposition?

Thermal radiation

28-28 The figure shows how the total power per unit area (intensity) of the incoming solar radiation varied on three cloudless days at a latitude of 52° N at different times of year. Estimate the total energy input per unit area on each day and work out roughly by what factor the total energy per unit area on June 25th exceeds that on January 6th.

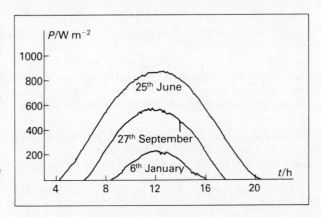

28-29 A mercury-in-glass thermometer with a bulb of surface area 1.2×10^{-4} m^2 was placed with its bulb at the centre of an empty saucepan; the thermometer initially read 20 °C and the saucepan was at a temperature of 80 °C. If the power losses per unit area from the surfaces of the thermometer and the saucepan are 260 W m^{-2} and 560 W m^{-2}

respectively, calculate the initial rate of gain of energy by the thermometer. [You can assume that the surface of the thermometer receives 560 W m^{-2}.]

If the heat capacity of the thermometer is 0.22 J K^{-1}, what temperature will it register after one minute?

28-30 The interior of a large furnace is at a temperature of 2800 K. What is the rate of emission of radiant energy from the open furnace door if this measures (a) 0.40 m by 0.60 m (b) 20 mm by 30 mm?

28-31 The graph shows the spectral distribution of power per unit area radiated by a black body at 10°C (283 K). Use the graph to estimate the total power per unit area radiated at this temperature and hence deduce a value for the Stefan-Boltzmann constant σ.

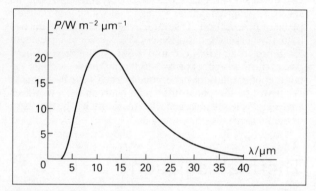

28-32 A baby, with an effective surface area of 0.21 m^2, is held by his mother above a bath. If the baby's skin temperature is 34°C at what rate does he radiate energy? Assume that his skin has a total emissivity of 0.55, i.e. it emits at 0.55 times the rate of a black body at the same temperature.

28-33 A black body is radiating at 800 K and, as a result, is cooling down at 0.40 K s^{-1}. After how long will it be radiating at 99% of its initial power output, i.e. in how long will its power output fall by 1%?

28-34 The element of a 1 kW electric fire has a surface area of 6×10^{-3} m^2. Estimate its working temperature. Explain what assumptions you have made in your estimation and hence explain whether your estimate is likely to be too large or too small.

28-35 The filament of a tungsten electric light bulb is raised to various temperates T. At each temperature the wavelength λ_m at which there is maximum radiation is measured. The results are as follows:

T/K	2400	1900	1450	1250	1050	900	850	750
λ_m/μm	1.8	2.1	2.5	3.3	3.8	4.7	5.7	6.5

Plot a graph to verify that the product $\lambda_m T$ is constant and hence find the value of the constant. Why does it not equal 2.9×10^{-3} m K, the constant in Wien's law?

28-36 A sphere (which may be assumed to be a black body) is maintained at a temperature of 300 K and suspended at the centre of a box which is effectively at 0 K. The sphere is found to radiate at a power of 100 W. Explain what the net

rate of loss of energy would be if each of the following changes was separately made:
(a) the temperature of the sphere was raised to 600 K
(b) the temperature of the enclosure was raised to 300 K
(c) the radius of the sphere was halved.

28-37 The baby in question **28-32** will also be absorbing radiant energy from his surroundings. Assuming that he absorbs 0.55 of the energy falling on his skin estimate his net rate of loss of radiant energy when the surroundings are at 25 °C.

28-38 A blackened copper cylinder of surface area 3.0×10^{-3} m^2 is suspended by thin wires in an evacuated enclosure. The walls of the enclosure are maintained at 273 K and a small electric heater embedded in the cylinder keeps its temperature constant at 400 K. What is the power supplied to the heater? Assume the cylinder acts like a black body.

If the cylinder has a heat capacity of 65 J kg^{-1} calculate the initial rate of fall of temperature of the cylinder when the current to the heater is switched off.

Electromagnetic wave spectra

28-39 The figure represents energy levels for four states of the hydrogen atom. The transitions shown by the full lines give rise to the three longest-wavelength lines in the Balmer series (they all lie in the visible part of the spectrum) of wavelengths 656 nm, 486 nm and 434 nm. Show that the energy values given in the figure are consistent with these wavelengths.

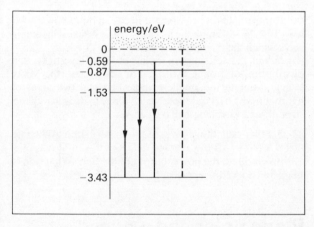

28-40 Calculate the longest wavelength radiation which might be emitted by a spectral transmission between any of the four energy levels shown in the figure for the previous question.

28-41 A narrow slit in a vertical board is placed in front of a low pressure mercury vapour lamp. A concave reflection grating with 400 lines per millimetre is placed 0.60 m from the slit, so that light from the slit is dispersed and focused by the grating on to flourescent paper fixed to the board.
(a) How far to one side of the slit will the first order spectral line of 254 nm wavelength ultra violet radiation appear?
(b) The second order spectral line for this ultra-violet line is amongst the first order visible spectral lines. How could you demonstrate simply the existence of the ultra-violet line?

28-42 The infra-red absorption spectra of many organic substances show sharp absorption lines at wavelengths of 3.2 μm and 8.2 μm. These are associated with the C—H bond. What characteristic bond energies correspond to these two lines? Give your answers in joules and in electron-volts.

28-43 Distinguish between the factors which give rise to the continuous X-ray spectrum and the characteristic line spectrum for a metal such as molybdenum which is often used as the target material in an X-ray tube used for radiography.

28-44 'γ-ray spectra give us evidence about the atomic nucleus, ultra-violet and optical spectra evidence about the outer atom and infra-red spectra evidence about the nature of interatomic bonding.' Write brief notes on this statement, including such things as reference to the scale of the energy changes which take place in each region.

Appendix: Useful mathematics

Measurements and units (A1–A6)

1 The width and height of a rectangular television screen are quoted as 0.46 m by 0.36 m. Write down the maximum and minimum values possible for the width and height of the screen consistent with this statement. Hence calculate the maximum and minimum values of the area of the screen to 4 significant figures; then state the area of the screen giving the result to the number of significant figures that are justified by the data.

What is the total force with which the atmosphere pushes on the screen if atmosphere pressure is 100 kPa?

2 The dimensions of a wooden plank are given as: length 1.825 m, width 0.178 m, thickness 0.019 m. Calculate its volume, giving the result to the number of significant figures justified by the data.

3 The density ρ of a substance is defined by the equation $\rho = m/V$, where m is the mass of a volume V of the substance. What is the SI unit of density. Calculate the density of (a) air (b) mercury (c) copper from the data given below. Be careful to state the results to the correct number of significant figures. You may assume that 1 litre $= 1 \times 10^{-3}$ m^3.
(a) When a flask of volume 195 ml is evacuated its mass is reduced by 0.251 g.
(b) A quantity of mercury is poured into a measuring cylinder and is found to have a volume of 124 ml; its mass is found to be 1684 g.
(c) A copper wire of length 0.804 m has a diameter of 0.71 mm and a mass of 2.84 g.

4 The unit of electrical resistance is the ohm (Ω); this is defined by $1 \, \Omega = 1 \, \text{V A}^{-1}$. The volt (V) is the unit of potential difference, and is defined by $1 \, \text{V} = 1 \, \text{J C}^{-1}$. The coulomb (C) is the unit of electric charge, and is defined by $1 \, \text{C} = 1 \, \text{A s}$; also $1 \, \text{J} = 1 \, \text{N m}$, and $1 \, \text{N} = 1 \, \text{kg m s}^{-2}$. Write down the definition of the ohm in terms of the SI base units.

5 The figure illustrates the sequence of steps by which certain mechanical quantities are defined, relating them to the base quantities *mass, length* and *time*. Redraw this diagram filling in the boxes with the units used for the quantities in the figure. The unit of power is called the *watt* (W). Use your diagram to show that $1 \, \text{W} = 1 \, \text{kg m}^2 \text{ s}^{-3}$.

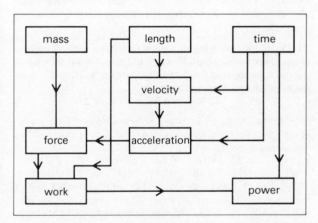

6 In an experiment to measure the speed c of electromagnetic radiation in a vacuum the wavelength of the radiation corresponding to the transition between the two hyperfine levels of the ground state of the caesium-133 atom is found to be 53 835.14 times as great as the wavelength λ_{Kr} of the radiation of the krypton-86 atom by which the metre is defined. The speed c is to be calculated from the relation $c = f \lambda$, where f and λ are the frequency and wavelength respectively of electromagnetic radiation in a vacuum. Calculate the value of c, given that for the specified radiation from caesium-133 the frequency f_{Cs} is given by
$$f_{\text{Cs}} = 9.192\,631\,770 \text{ GHz}$$
and that $1 \text{ m} = 1\,650\,763.73 \, \lambda_{\text{Kr}}$ (exactly)

7 Write down the following quantities as numbers in standard form (to 2 significant figures) together with the appropriate unit without any prefix (such as p or M).
(a) 800 pF (b) 1.5 kV (c) 50 MW (d) 40 ns.

Given that $1 F = 1 C V^{-1}$, $1 W = 1 J s^{-1}$ and $1 V = 1 J C^{-1}$, work out the following and express the results by means of numbers in standard form with a *single* unit:
(e) the product of (a) and (b)
(f) the product of (c) and (d)
(g) the result in (f) divided by the result in (e).

8 The cost of electrical energy is quoted as 2.85p per kilowatt-hour. Work out the cost of cooking a large meal, if this requires the use of three cooking rings for 2.5 hours; the rings are each rated at 2.8 kW.

9 In an old scientific textbook the intensity of the solar radiation arriving at the Earth (before absorption, etc by the atmosphere) is quoted as 2.0 cal per square centimetre per minute. Express this quantity in the appropriate SI unit, given that 1.0 cal = 4.2 J.

According to theory the relation between T and m is of the form $kT^2 = m + b$ where k and b are constants. Explain why this can be regarded as a *linear relation* between T^2 and m, and establish the point by plotting a graph of T^2 against m. Use your graph (a) to predict the period of oscillation to be expected for a mass of 50 g hanging on this spring (b) to find the intercepts on the two axes (c) to measure the gradient of the graph, and (d) from the results of (b) and (c) work out the values of b and k. [Be careful to state the units of these quantities correctly.]

15 A pendulum bob is pulled to one side and then released. Its speed v at different times t after release are given in the tables:

t/s	0	0.10	0.20	0.30
v/m s^{-1}	0	0.24	0.47	0.68

t/s	0.40	0.50	0.60	0.70
v/m s^{-1}	0.86	0.99	1.07	1.11

Plot a graph of v against t; draw a tangent to the curve at the point where $t = 0.50$ s, and measure the gradient (dv/dt) at this moment.

Graphs (B1–B3)

10 'Provided the force F acting on a body stays constant, the increase in its velocity v is equal to the product of a constant acceleration a and the interval of time t for which the force acts.' Express this sentence using only mathematical symbols.

11 Write the following statement out in full without mathematical symbols, where V refers to the potential difference across a thermistor, I refers to the current in it, and T is its thermodynamic temperature:
$$\Delta V = -\,2\ V, \text{ when } \Delta T = 10\ K, \text{ if } \Delta I = 0$$

12 The following table gives a series of values of the wavelength λ of sound in air measured for different values of the frequency f.

f/kHz	0.20	0.40	0.60	1.00	1.50	2.00
λ/m	1.71	0.86	0.57	0.34	0.23	0.17

Plot graphs (a) of f against λ and (b) of f against $1/\lambda$, with f on the vertical axis in each case. What can you deduce from these graphs about the relation between f and λ? Measure the gradient of whichever graph is a straight line (bearing in mind that 1 Hz = 1 s^{-1}).

13 The potential difference V across an electric cell is measured for different values of the current I taken from it, and the following results are obtained:

I/A	0	0.20	0.40	0.60	0.80	1.00
V/V	1.52	1.40	1.28	1.16	1.04	0.92

Plot a graph of V (vertically) against I. and measure its gradient. Write down the equation connecting V and I for this cell.

14 The following tables gives the period T of vertical oscillations of a mass m hung on the end of a spring.

m/kg	0.100	0.200	0.300	0.400	0.500
T/s	1.12	1.43	1.68	1.90	2.10

Angles (B4–B5)

16 An arc of a circle of radius 2.50 m subtends an angle of 60.0° at the centre. Find (a) the above angle in radians (b) the length of the arc.

17 A motor car tyre of radius 0.270 m is punctured in close sucession by two tin-tacks on the road. The angle subtended at the centre of the wheel by the part of its circumference between the two tacks is 140°. What was the shortest possible distance on the road between the tin-tacks?

18 The visible diameter of the planet Venus is 1.22×10^7 m, and its closest distance of approach to the Earth is 4.1×10^{10} m. Calculate its maximum angular size as seen from the Earth (a) in radians (b) in minutes of angle.

19 A unit of distance used in astronomy is the *parsec;* this is defined as the distance from the solar system at which the radius of the Earth's orbit (1.50×10^{11} m) subtends an angle of 1 second. Calculate the size of the *parsec* in metres.

The nearest star to the solar system is α *Centauri* at a distance of 1.32 parsec. What angle does the radius of the Earth's orbit subtend at α *Centauri*?

20 The top of a tree is seen at an elevation of 9.0° above its base when observed from a point at a distance of 120 m from it. Estimate the height of the tree.

21 In order to measure the angular width of the scene photographed through a camera lens a boy takes a picture of a horizontal metre scale set at right-angles to the lens axis and at a distance of 1.2 m from it. The developed photograph shows 0.92 m of the scale. Calculate the angular width of the scene (a) assuming that 0.92 m is the length of the circular arc between the extreme edges of the scene (b) taking this distance more exactly as the shortest distance between these edges. What is the percentage error in using the approximate method of calculation?

Sinusoidal oscillations (B6)

22 When a tuning fork is set in vibration the displacement x of the end of one of the prongs from its equilibrium position at time t is given by $x = x_0 \sin 2\pi ft$. The frequency f of a particlar fork is 440 Hz. Calculate the maximum speed v_0 of the ends of its prongs, if the maximum displacement x_0 is 0.50 mm. [The speed v is equal to dx/dt.]

If the speed v of the ends of the prongs also varies sinusoidally, calculate the maximum acceleration ($= dv/dt$) of the prongs.

23 At a certain point in a radio set tuned to the BBC long-wave transmitter there is a sinusoidal alternating potential V of peak value 2.5 V and frequency 200 kHz. Calculate (a) the maximum rate of change of V (b) the value of V 0.20 μs after it changes direction.

24 The tides of the sea move up and down a certain beach at a frequency of approximately twice per day. What is this frequency in Hz [Note: 1 Hz = 1 s^{-1}]? If the amplitude of movement of the tide is 15 m on either side of the mean tide-line, what is the maximum rate at which the tide moves up the beach, assuming that the variation is sinusoidal?

Exponentials (B7)

25 Plot the exponential curve $z = z_0 e^{bt}$ for values of t at 2 s intervals from 0 up to 10 s, taking $b = 0.2$ s^{-1} and $z_o = 2$.

Measure the gradient of the above curve (dz/dt) at a number of points by drawing tangents at these points; and plot another graph of dz/dt against z. Test whether the value of b deduced from this graph agrees with that used above.

26 A certain bacterial population number N grows exponentially in the right environment according to the relation
$$N = N_0 e^{bt}$$
Where N_0 is the number of bacteria when $t = 0$. In an environment of tinned meat at room temperature the population doubles in 5 hours. What is the value of b for this population? If a tin of meat is polluted with a single live bacterium, what will the bacterial population become after one week?

27 The excess pressure p in a leaky gas cylinder is measured at 9 am every morning at various times t (in days) with the following results:

t/d	0	1	2	3	4	5
p/MPa	6.0	3.8	2.4	1.5	0.9	0.7

Plot a graph of $\ln(p/\text{MPa})$ against t. Do you consider that the measurements justify assuming that the pressure excess in the cylinder is decaying exponentially? If so, calculate the time constant of the decay process. At what moment would you expect the pressure to have fallen to exactly half its initial value?

Units and symbols

physical quantity	SI unit	symbol
length	metre	m
mass	kilogram	kg
time	second	s
electric current	ampere	A
thermodynamic temperature	kelvin	K
amount of substance	mole	mol

Table 1 Names and symbols for SI base units

physical quantity	SI unit	symbol	SI equivalent
frequency	hertz	Hz	s^{-1}
force	newton	N	$kg\,m\,s^{-2}$
energy	joule	J	$N\,m$
power	watt	W	$J\,s^{-1}$
pressure or stress	pascal	Pa	$N\,m^{-2}$
electric charge	coulomb	C	$A\,s$
electrical p.d. or e.m.f.	volt	V	$J\,C^{-1}$
electrical resistance	ohm	Ω	$V\,A^{-1}$
electrical capacitance	farad	F	$C\,V^{-1}$
magnetic flux density	tesla	T	$N\,A^{-1}\,m^{-1}$
magnetic flux	weber	Wb	$V\,s$
inductance	henry	H	VsA^{-1}

Table 2 Special names and symbols for some SI derived units

physical quantity	unit	symbol	value
mass	tonne	t	$10^3\,kg$
volume	litre	l	$10^{-3}\,m^3$
time	hour	h	$3600\,s$
angle	radian	rad	π rad = 180°
energy	kilowatt hour	kW h	3.6×10^6 J
energy	electron-volt	eV	1.60×10^{-19} J
temperature	degree Celsius	°C	0°C = 273.15 K
radioactivity	curie	Ci	$3.7 \times 10^{10}\,s^{-1}$

Table 3 Non-SI units. These units are commonly used together with those in table 2 in expressing the value of a physical quantity

multiple	prefix	symbol	example
10^{-12}	pico	p	pF
10^{-9}	nano	n	nm
10^{-6}	micro	μ	μA
10^{-3}	milli	m	mV
10^3	kilo	k	kPa
10^6	mega	M	MΩ
10^9	giga	G	GHz

Table 4 SI prefixes

Answers

1 Describing motion

1-2 8.0 m s^{-1}, 1.6%

1-3 (a) 35 ms (b) 2.8%

1-8 (a) $+9.0$ m s^{-1} (b) -21 m s^{-1}
(c) -12 m s^{-1} (d) 20 m s^{-1} west

1-10 (a) 10 m s^{-1} (b) 10 m s^{-1} (c) 0
(d) 6.4 m s^{-1} south (e) 9.0 m s^{-1} SE
(f) 20 m s^{-1} west (g) 14 m s^{-1} SW

1-11 (a) 8.4 m, N $17°$ W
(b) 2.9 m s^{-1}, S $79°$ W

1-12 (a) 3.0 m north (b) 112 m N $27°$ E
(c) 0 (d) 127 m north

1-13 (a) 3.2 m s^{-1}, N $39°$ W (b) 20 s
(c) 40 m

1-14 (a) $37°$ (b) 1.5 m s^{-1} (c) 33 s

1-15 9.2 m s^{-1} south

1-16 (a) 10.4 m, N $48°$ E (b) 8.7 m s^{-1},
S $14°$ E

1-17 125 m s^{-1}, N $61°$W

1-18 -0.89 m s^{-2}

1-19 5.0 m s^{-2}

1-20 1.8 km, 32 s

1-21 (a) $+4.4$ m (b) -8.1 m

1-22 after 2.0 s

1-23 400 m

1-24 (a) 0.15 km (b) 0.49 km

1-25 (a) 0.93 m s^{-2} (b) 3.0 m s^{-2}

1-31 (a) in the same place
(b) 80 mm further back

1-32 (a) 0.45 s (b) 0.64 s

1-33 10.1 m s^{-2}

1-34 1.2 m, $1.4°$, a greater amount

1-35 (a) 12.5 m s^{-1} (b) 1.3 s (c) 8.0 m
(d) 28 m

1-36 $\sqrt{s}/\text{m}^{1/2} = 0.22, 0.36, 0.51, 0.66, 0.81$

1-37 slope $= 2.25$ m$^{1/2}$ s^{-1}, $g = 10.1$ m s^{-2}

1-38 $40°$ below the horizontal

1-39 9.8 m s^{-2}

2 Momentum and force

2-1 (a) (i) 150 kg m s^{-1} north,
(ii) 150 kg m s^{-1} south
(b) 2.0×10^4 kg m s^{-1} east
(c) 5.0×10^9 kg m s^{-1} west

2-2 0.28 kg, (a) yes (b) yes

2-3 0.79

2-4 3.0 m s^{-1} in the other direction
(b) 0.52 m s^{-1} south

2-5 $+1.7$ m s^{-1}

2-6 3.40×10^5 m s^{-1}

2-10 24 m s^{-1}

2-15 (a) 10 N (b) 17 N (c) 17 N

2-18 (a) yes (b) 5.6 N, 2.1 N

2-19 (a) 31 N (b) 60 N

2-20 740 N, N $22°$ E

2-23 (a) 20 N (b) 12 N

2-25 Yes: downwards

2-26 (a) 12 kN (b) 12 kN

2-27 1.5

2-28 20 kN

2-29 (a) 0.55 m s^{-2} (b) 1.2 kN

2-30 $3.1°$

2-31 15 m s^{-1}, 14 kN

2-32 (a) 4.6 m s^{-2} (b) 0.81 m s^{-2}

2-33 4.9 m s^{-2}

2-34 (a) 2.8 kN (b) 1.6 kN

2-37 thread $4.7°$ to the vertical

2-38 (b) 7.8 N (c) 3.9 kg

2-40 zero

2-42 (a) 3000 N s east
(b) 100 N s downwards
(c) 100 N s downwards

2-43 (a) 2.7 N s away from the player
(b) the same

2-44 (a) 32 N (b) 6.4 N

2-45 2.5 N s, 42 m s^{-1}, 1.5×10^4 m s^{-2}

2-46 0.68 m s^{-1}

2-47 7.4 kN, a small unladen car

2-48 (a) 0.40 m s^{-1} to the left (c) ± 24 kN

2-54 (a) 6.0 kN (b) 5.2 kN

2-56 440 N

2-57 (a) 4.8 m s^{-2} (b) 0.29 kN (c) 0.29 kN

2-58 (a) $a = 0$ (b) $a = 0$ (c) $a = 2.0$ m s^{-2}

2-60 5.2 kN

2-61 1.7 N

2-62 8.9 m s^{-1}, 2.7 m s^{-2}

2-63 1400 m s^{-1}, 1000 kg s^{-1}, 3.4×10^7 N

2-64 120 m s^{-1}

2-65 170 kN, the air which is being accelerated

3 Energy and its conservation

3-1 $0, 0, 24$ J, -16 J

3-2 0.59 N, (a) 0, 12 mJ (b) 0, -12 mJ

3-3 330 kg, 20 J, 20 J, a force-multiplier

3-4 none

3-5 (a) $2\pi r(F_1 - F_2)$ (b) 53 W

3-6 60 W to 100 W

3-7 (a) 0.80 J (b) 2.4 J

3-8 2.6 J, 60 mJ

3-9 54 mJ

3-10 -5.1×10^{10} J

3-11 98 W

3-12 50 kN

3-13 (a) ≈ 25 J (b) ≈ 450 J (c) ≈ 3.5 kJ
(d) ≈ 20 MJ

3-14 (a) 1.6 kN (b) 800 m

3-15 (a) 7.4 J (b) 7.0 m s^{-1} (c) -0.41 J

3-16 (a) g.p.e. -13.7 kJ, k.e. $+13.7$ kJ
(b) g.p.e. -2.7 kJ, k.e. -13.7 kJ,
e.p.e. $+16.4$ kJ; force $= 4.1$ kN

3-17 0.83 m s^{-1}, no

3-18 (a) 112 J (b) -75 J
(c) 4.4 m s^{-1} (d) -112 J (e) 22 N

3-19 0.30 kN

3-20 17 m s^{-1}

3-21 2.3 m s^{-1}

3-22 (a) 6.3×10^{-13} J (b) 1.4×10^7 m s^{-1}

3-23 (b) 1.2 J (c) -0.60 J (d) 0.60 J
(e) 1.5 m s^{-1} (f) -0.60 J
(g) varying from 4.9 N to zero

3-24 98%

3-25 (a) 0.25 J (b) 0.049 J

3-27 (a) $2\cdot1$ m s^{-1}, $1\cdot06$ m (first mid point)
(b) k.e.$0\cdot21$ J, g.p.e. $1\cdot04$ J

3-28 (a) 0.20 J
(b) 0.10 J; the push of the man's hand
does negative work

3-29 a ball bouncing on the ground.
Possible values of m and h are 0.40 kg
and 1.0 m

3-30 0.12 J, 8.9 m s^{-1}

3-31 (a) -0.80 J (b) 0.40 J (c) 0.40 J
(d) -1.6 J (e) 1.6 J (f) the lump has
no k.e.

3-33 13 N m^{-1}

3-34 (a) 16 N (b) 16 N, 4.0 N, $F \propto 1/r^2$

3-35 (a) about 10 MJ (b) about 2 kW

3-38 43 MJ, 36 p

3-40 1.7×10^7 m^2, 0.007%, 0.3%

3-41 640 MW

3-42 1.05 GW, 4.8×10^9 kg

3-43 2.2 m s^{-1} north, 1.5×10^5 J

3-44 no

3-45 (a) 0, 0.30 m s^{-1}
(b) -0.10 m s^{-1}, 0.20 m s^{-1}

3-46 (a) 0.40 km s^{-1} (b) 0.79 kJ

3-49 70 g

3-50 6.0 m s^{-1}, N $68°$ E, 2.8×10^4 J

3-51 the particle of mass m

3-52 (a) $-u$, $u \times 10^{-5}$
(b) u, $2u$. Unchanged

3-53 (a) $0.14c$ (b) $0.42c$ (c) $0.75c$

3-54 9.10×10^{-31} kg, 1.67×10^{-27} kg

3-55 (a) 3.1×10^{-12} kg (b) 1.7×10^{-10}

3-56 (a) $4.76\,493 \times 10^{-29}$ kg
(b) $7.118\,540 \times 10^{-3}$ (c) 26.8 MeV

3-57 the mass increases

4 Structure of matter

4-1 (a) 7.00, 28.00 (b) yes: NO, N_4O,
N_2O
(c) yes: NO_2, N_2O, NO
(d) 14, because it gives the simplest
formulae

4-2 (a) 1.5×10^{-10} m^3 (b) 6.6×10^{-2} m^2
(c) 2.3×10^{-9} m (d) 1.3×10^{-10} m

4-3 (a) about 2000 (b) about $30\,000$

4-4 2.1×10^{-4}, 2.4×10^{17} kg m^{-3}

4-5 $m_H : m_e = 1840 : 1$

4-6 (a) 12, 13, 14 (b) 2.3×10^{-26} kg

4-7 28.1

4-8 (a) $A_r = 39$
(b) 1.24×10^4 m s^{-1}, 1.21×10^4 m s^{-1}
(c) 48.3 μs, 49.5 μs

4-9 (a) 5.5×10^{-7} kg (b) 0.32 kg (c) 0.32
kg

4-10 (a) 9.3×10^{-2} mol (b) 18 mol

4-11 (a) 6.0×10^{24}

4-12 256, 4.25×10^{-25} kg

4-13 (a) 5.90×10^{28} (b) 2.6×10^{-10} m
(c) 6.04×10^{28}, 2.55×10^{-10} m

4-14 (a) 4.44×10^{-29} m^3
(b) 9.59×10^{-26} kg
(c) 6.10×10^{23} mol^{-1}

4-20 (a) 23.8×10^{-9} N, 6.9×10^{-9} N,
16.9×10^{-9} N (repulsive)
(b) 1.3×10^{-9} N, 3.9×10^{-9} N,
2.6×10^{-9} N (attractive)

4-21 (b) about 0.15×10^{-10} m (c) no
(d) 7.3×0^{-9} N, zero

4-27 about 6 days

4-28 $300d^3$, $7d$

5 Performance of materials

5-1 (a) 24 MPa (b) 0.16 GPa (c) 11 MPa
5-2 (a) 5.0×10^{-5} (b) 2.0
5-3 (a) 59 N (b) 59 N (c) 59 N (d) 69 N
  (e) 64 N
5-4 (a) 50 N, 50 N (b) 50 MPa, 0.10 GPa
5-5 end-to-end: force
  side-by-side: extension, strain, stress
5-6 (a) 8.9 m s^{-1} (b) 14 kN (c) 28 kN
  (d) 47 MPa
5-7 0.51 mm, 1.6 mm
5-8 (a) 37 kN (b) 12 MPa (c) 5.9×10^{-5}
5-9 (a) 0.10 GPa (b) 7.7×10^{-4}
  (c) 0.92 mm (d) 4.6 mJ
5-10 (a) tungsten (b) copper
5-12 (a) high tensile steel (b) 2.5 J, 8.1 J
  (c) mild steel
5-13 aluminium
5-14 all true
5-16 (a) 30 GPa (b) 3.0 mm (c) 65 mJ
  (d) 32 mJ
5-23 (a) 8.47×10^{28}, $d = 2.28 \times 10^{-10}$ m
  (b) 1.93×10^{13} (c) 20 MPa
  (d) 1.5×10^{-4} (e) 3.5×10^{-14} m
  (f) 1.0×10^{-12} N (g) 30 N m^{-1}
5-25 (a) P/mnd^2 (b) $\Delta d/d$
  (c) $E = P/mnd\Delta d$ (d) P/mn
  (e) $\Delta d = P/mnk$ (f) $k = Ed$ (g) 31 N m^{-1}

6 Fluid behaviour

6-1 1.1 kPa, 1.6 kPa, 3.3 kPa
6-3 about 7 kPa
6-4 0.43 kN
6-5 8.0 km
6-6 (a) B (b) 103.0 kPa
6-7 about 1·1 m
6-9 (a) 50 kPa (b) 0.50 kN (c) no
6-11 (a,b,c) 2.6×10^{-2} N, 0.15 m
6-12 (a) 1.6×10^6 N (b) 1.5×10^6 N
6-13 2.0 kN
6-14 (a) 0.15 N (b) 0 (c) 0.024 N
6-15 (a) 2×10^5 N (b) 7×10^4 N (c) 60 m^3
6-18 (a) 0.80 m s^{-1}
  (b) (i) $8.0 \times 10^{-5} \text{ m}^3 \text{ s}^{-1}$
  (ii) $8.0 \times 10^{-5} \text{ m}^3 \text{ s}^{-1}$
6-19 (a) 0.98 kPa (b) 2.0 (c) 0.81 m s^{-1}
  (d) $2.5 \times 10^{-2} \text{ m}^3 \text{ s}^{-1}$
6-23 (a) 28 mm s^{-1} (b) 0.14 m s^{-1}
6-24 1.2 N s m^{-2}
6-26 22 mm s^{-1}
6-27 $8.0 \times 10^{-2} \text{ N s m}^{-2}$
6-28 $T \propto \sqrt{(l/g)}$
6-29 $T \propto \sqrt{(m/k)}$
6-30 (a) (c) (d) and (f)
6-31 0.28 mJ
6-32 140
6-33 (a) 75 m s^{-2} (b) 75 m s^{-2}
6-34 (a) 24 mm (b) 48 mm, 46 mm
6-35 (a) 0.29 kPa (b) 1.9 kPa (c) 24 Pa
6-36 20 Pa, 8.0 Pa, 12 Pa, 10 mm
6-37 0.47 N m^{-1}
6-38 $h/\text{mm} = +92, +31, -31$
  (a) 100.5 kPa (b) 101.5 kPa (c) 0.15 mm
6-41 $T \propto \sqrt{(\rho a^3/\gamma)}$

6-42 both possible

7 Electric circuits

7-2 electric, pull together
  magnetic, push apart
7-3 B 0.78 A, C 0.39 A
7-4 $I_B = 0.05$ A, not bright
  $I_C = 0.20$ A, bright
7-5 5.0 A
7-6 1.8×10^2 C, 1.1×10^{21}
7-7 (a) 0.26 A (b) 4.7 kC (c) charge
  (d) 6.3 kC
7-8 $3.4 \times 10^{15} \text{ s}^{-1}$
7-9 7.8×10^7
7-10 (a) $2.2 \times 10^{-4} \text{ m}^3$
  (b) 2.9×10^{-11} C (c) 4.8×10^{-14} A
7-11 0.75 mm s^{-1}
7-12 1:16
7-13 $9.4 \times 10^{20} \text{ m}^{-3}$
7-14 $5.07 \times 10^6 \text{ C kg}^{-1}$
  $2.01 \times 10^6 \text{ C kg}^{-1}$
7-15 2.5×10^{19}, 0.40 A
7-16 $8.92 \times 10^5 \text{ C kg}^{-1}$
  1.60×10^{-19} C
7-17 5.6 A
7-18 3.0 V
7-19 36 kJ, 25 h
7-20 0.24 kJ (a) 0.72 kJ (b) 0.24 kJ
7-21 80 C, 1.3 A
7-22 8.0×10^8 V, about 1×10^4
7-23 (a) 3.0×10^6 C
  (b) 2.4×10^7 J, internal energy
7-24 (a) A +6.0 V, B +3.5 V
  (b) A 0, B -2.5 V; 3.5 V, 0.70 kJ
7-25 (a) 1.35 V (b) 0.45 V (c) 0.15 V
  (d) -0.30 V (e) 1.35 V
7-26 R_1 0.20 A; R_3 R_4 0.30 A, cell 0.50 A
  (a) R_1 R_2 4.0 C, R_3 R_4 R_5 6.0 C, cell
  10 C
  (b) R_1 3.0 J, R_2 2.4 J, R_3 R_4 R_5, 2.7 J
  each (c) 15.0 J (d) 0.15 V (e) 1.5 J
7-27 0, 1.2 V, 2.4 V, 0, 1.2 V, 1.8 V;
  no current in BE
7-30 (a) A +1.4 V, B -1.4 V, D 0,
  CD makes no effect
  (b) A B D -1.4 V, no current
  CD makes potential of D zero, and
  currents flow
7-31 (a) both resistors 0.40 A, CD zero
  (b) both resistors 0.40 A, CD 0.80 A
7-32 (a) 0.50 V m^{-1} (b) 30 V m^{-1}
  (c) $2.0 \times 10^4 \text{ V m}^{-1}$ (d) 200 V m^{-1}
7-33 copper (a) 3.1 g
7-34 (a) 6.0 V (b) 1.5 V (c) 3.0 V
  (d) 4.5 V (e) 3.0 V; (b) and (c)
7-35 5 cells; 3.6 V
7-36 50
7-37 15 W
7-38 30 W, 3.6 kJ
7-39 5.8 V
7-40 18 kJ, 0.25 A
7-41 200 MW, 1.1×10^{12} J h^{-1}
7-42 500 A (a) 120 V (b) 60 kW
7-43 29 A
7-44 20 ms (a) 5.0 ms (b) 3.3 ms
7-45 12.5 rev s^{-1}

8 Electrical resistance

8-1 25 Ω
8-2 8.6×10^{10} Ω
8-3 (a) 0.040 Ω^{-1} (b) $1.2 \times 10^{-11} \text{ Ω}^{-1}$
8-4 50 V
8-5 (a) 99 kΩ (b) 83 kΩ; about 200 V
8-6 (a) 20 kΩ (b) 3.0 V, 0.15 mA
  (c) 4.0 V
  (d) first 0.20 mA, second 0.10 mA
  (e) 0.20 mA
8-7 both 3.0 V, 0.30 mA, R_1 2.5 V,
  0.25 mA; R_2 3.5 V, 0.35 mA, 0.10 mA
  25 kΩ
8-8 (a) 0.143 mA (b) 0.333 mA
  (c) 0.190 mA (d) 5.25 kΩ
8-9 (a) 0.31 A (b) (i) 11 Ω (ii) 15 Ω
  (c) 2.7 V (d) 8.0 V
8-10 (a) 5 Ω (b) 2.0 Ω (c) 0.75 Ω, 0.37 A
8-11 4.0 V. 1.6 J s^{-1}
8-12 0.25 mA, 1.25 W
8-13 (a) 0.96 kΩ (b) 0.38 kΩ (c) 2.4 Ω
  (d) 20 (e) only the 240 V bulb bright
8-14 (a) 33 A (b) 6.0 kV, 5.9 kV, 0.10 kV
  (c) 5.8 kV (d) 6.7 kW (e) 3.3% (f) 30%
8-15 (a) 5.0 Ω (b) 1.2 Ω
8-16 (c) < 1 Ω
8-17 6
8-18 (a) 2.35 Ω (b) 5.00 Ω (c) 7.40 Ω
  (d) 4.80 Ω
8-19 (a) 6.0 V (b) 3.0 V (c) 1.9 V
  (d) 3.0 V; (b) 0 (c) -4.1 V
8-20 12.0 m
8-21 3.0 V, 2.0 V
8-22 2.40 V, 1.28 V
8-23 (a) 4.0 V (b) 4.8 V (c) 0
8-24 0.9, 1.6, 2.1, 2.4, 2.5, . . .
8-25 (a) 1.5 V (b) 1.5×10^{-4} A
8-26 1.00 kΩ
8-27 (a) 1.72 Ω in parallel
  (b) 830 Ω in series
8-28 (a) (iv) 0.012 8 Ω in parallel
  (b) (i) 18.75 kΩ in series
8-29 4.4×10^{-3} Ω in parallel
8-30 (a) 0.075 Ω in parallel
  (b) 66.2 kΩ in series
8-31 (a) 14 kΩ (b) 84 µA (c) 1:14
  (d) 5.6 µA (e) $4.0 \times 10^{-8} \text{ A div}^{-1}$
8-32 4.8 Ω, from B to D
8-33 0, 24.4 mV; D, AB 400 Ω
8-34 8.16 Ω, 625 mm from A
8-35 (a) 2.0 V, 1.0 V, 1.0 V (b) 0
  (c) all 1.0 mA (d) P 1.0 mA, Q 2.0 mA
8-36 0.75 V, 1.50 V, from D to B
8-37 0.50 Ω (a) 0.60 J s^{-1} (b) 0.08 J s^{-1}
8-38 (a) 1.52 V, 0.64 Ω (b) 1.04 V, 0.60 Ω
8-39 (a) 3 A (b) 200 A
8-40 (a) 3.0 V, 0.8 Ω, 1.25 A
  (b) 1.5 V, 0.2 Ω, 0.83 A
8-41 (a) 2.4 Ω (b) 0.15 Ω (c) 0.60 Ω
  (d) 1.5 Ω (e) 0.80 Ω
8-42 0.20 A, 1.4 V (a) 42 J (b) 3 J
8-43 (a) 8.3 Ω (b) 11.7 Ω (c) 1.7 Ω
8-44 7.5 V, 1.0 Ω, 7.35 V
8-45 8.5 Ω (a) 0.16 kW (b) 6 W + 34 W
8-46 6Ω; 6 V, 6 W
8-47 15 V, 2.1 Ω

8-48 (a) 9.0 A (b) 2.0 A (c) 29 V; 14 V
8-49 (b) 30 A (c) 90 A (d) 1.0 Ω
(e) 7.0 V, 5 W
8-50 (a) 1.2 V cell, 8.0 Ω in series
(b) diode (resistance 6.7 Ω at 2.0 V) and 20 Ω in parallel
8-51 1.9 Ω, 0.24 V m⁻¹
8-52 2.0×10^{15} Ω m, 5.0×10^{-16} Ω⁻¹ m⁻¹
8-53 5.5 m
8-54 (a) 1.26×10^{-5} m² (b) 2.5 Ω
(c) 0.42 Ω km⁻¹
8-55 $R = \rho/d$, 80 Ω
8-56 8.2
8-57 1.7×10^{-4} K⁻¹
8-58 2 Ω, 12 times; no (T increases about 8 times)
8-59 (a) 8.8 V (b) 15 V (c) 10 V (d) 0.36 A
(e) 0.96 A
8-60 3.7×10^{-4} m s⁻¹
3.7×10^{-1} m² s⁻¹ V⁻¹
8-61 (a) $(4l^2\rho l)/(\pi d^2)$ (b) πdl
(c) $(4l^2\rho)/(\pi^2 d^3)$; 4.4×10^5 W m⁻²
8-62 1.39 V
8-63 both 1.018 V
8-64 125.6 Ω
8-65 −0.14 A
8-67 0.75 m, 0.29 m, 0.94 m
8-68 3.0 mV
8-71 498 Ω, 4 mm

9 Heating solids and liquids

9-3 (a) heating: i.e. from can, i.e. to room
(b) working: c.e. from man, i.e. to wood
(c) heating: i.e. from heater, i.e. to room
(d) working: c.e. from boy, i.e. to pump
(e) working: g.p.e. from ball, i.e. to it and surroundings
(f) working: k.e. from coffee, i.e. to coffee
9-4 (a) +0.20 MJ (b) −0.080 MJ
(c) +0.12 MJ
9-7 (a) 0.19 m (b) 17 cm³
9-8 2.3×10^{-5} K⁻¹
9-9 93 mm
9-10 150.18 mm, 120.14 mm, 27 mm²
9-11 4.3 mm (a) 0.30 m (b) 60 MPa
(c) 60 MPa
9-12 0.17%
9-13 3.65×10^{-3} K⁻¹
9-14 747 kg m⁻³
9-19 8.4 kJ K⁻¹ (a) 8.4 kJ K⁻¹
(b) 0.31 kJ K⁻¹ (c) 38 J K⁻¹
(d) 14 J K⁻¹
9-20 7.0 kW
9-21 26 p
9-22 about 5 kJ
9-23 0.12 K s⁻¹
9-24 about 5 min
9-25 100 K
9-27 91°C
9-28 0.38 kJ K⁻¹ kg⁻¹
9-30 240 s (a) 542 W (b) 278 W (c) 66 kJ
9-31 23.2°C, 0.99 kJ K⁻¹ kg⁻¹
9-32 5.0 W
9-33 (a) 3.57 K min⁻¹, 2.34 K min⁻¹

(b) 67.8 K, 44.5 K (c) yes (d) 36 W
9-34 (a) 30 W (b) 10 W 2.0 K min⁻¹
9-35 (a) 0.50 kW (b) 0.50 kW (c) 6.0 g s⁻¹
9-36 (a) 9.5 kJ (b) 8.0 kJ (c) 16 W
(d) 5.1×10^2 s
9-37 (a) 0.18 kJ K⁻¹ kg⁻¹
(b) overestimate (c) low s.h.c. (d) no
9-38 0.43 kJ K⁻¹ kg⁻¹, 0.41 kJ K⁻¹ kg⁻¹
9-40 5.0 kJ K⁻¹ kg⁻¹
9-42 8.2 V
9-43 0.53 kJ K⁻¹ kg⁻¹, 0.88 kJ K⁻¹ kg⁻¹
9-44 24.7 J K⁻¹ mol⁻¹
9-45 (a) 0.50 kJ (b) 0.50 kJ (c) 0.25 kJ
(d) 50 J (e) 50 J
9-46 (b) (i) 7.1 kJ, (ii) 11.5 kJ
9-47 (a) 4.5 MJ (b) 0.17 MJ
9-48 (a) 38 min (b) 74 min
9-51 0.36 kg
9-52 24 g
9-53 14 g
9-54 15.4 W
9-55 23 min
9-56 1.28 MJ kg⁻¹, 1.14 MJ kg⁻¹. 2.72 W
9-57 (a) 42 MJ (b) 11 MJ (c) 53 MJ
(d) 20%
9-59 1.8 K min⁻¹, 4.8 K min⁻¹
(a) 0.69 kJ K⁻¹ kg⁻¹ (b) 14 W
(c) 87 kJ kg⁻¹
9-60 (a) at 70°C, rate of loss of energy = 20 W
(b) 0.24 MJ kg⁻¹
9-62 (a) 1.25×10^{-4} m³
(b) 3.72×10^{-2} m³ (c) 3.74 kJ
(d) 39.4 kJ (f) 9.5%
9-66 (a) 55 W (b) 28 W (c) 14 W (d) 11 W
9-67 0.47 kW
9-68 72°C, 57°C
9-69 (a) 1.04 kg s⁻¹ (b) 3.5 kW
(c) 2.3 W m⁻¹ K⁻¹
9-70 1.0 kW, 23°C, 60 W
9-71 (a) 40 V (b) R (c) 30 V (d) 40°C
(e) l (f) 30°C
9-73 (a) 18 W (b) 18 W
(c) 5.5×10^2 K m⁻¹ (d) 2.7×10^2 K m⁻¹
9-75 395 W m⁻¹ K⁻¹
9-77 0.33 W m⁻¹ K⁻¹
9-78 0.10 W m⁻¹ K⁻¹
9-79 (a) W m⁻² K⁻¹
(c) U/W m⁻² K⁻¹ = 8.3, 3.3, 0.63, 250,
10. R/m² K W⁻¹ = 0.12, 0.30, 1.60,
0.0040, 0.10
(d) R/m² K W⁻¹ = 0.32, 2.04, 0.204
(e) U/W m⁻² #⁻¹ = 3.1, 0.49, 4.9
(f) (i) 0.84 kW, (ii) 0.13 kW (g) 0.18 kW
9-80 0.36 K, 10.8°C, 11.2°C
9-81 no: U = 1.32 W m⁻² K⁻¹
9-82 0.14 W m⁻² K⁻¹
9-88 80 K temperature rise
9-89 (a) 4 (b) 0.5

10 Thermometers and the ideal gas

10-2 (a) 19.4°C (b) 19.0°C
10-5 (a) low (b) low
10-6 505.1 K

10-7 629.976 K, 629.923 K, 629.863 K
629.81 K
10-8 (a) 290.00 K, 289.77 K
(b) 16.85°C, 16.62°C
10-11 49.07°C
10-13 40°C
10-14 (a) 0.18 m (b) 32 mm
10-15 (a) 80 kPa (b) 0.14 MPa
(c) 0.15 MPa (d) 60 kPa
10-16 (a) 85 kPa (b) 0.16 MPa
(c) 0.16 MPa (d) 0.18 MPa (e) 70 kPa
10-17 1.7×10^{-4} m³
10-18 2.1 mol (a) 4.1 g (b) 66 g
10-19 greatest (a), least (b)
10-20 about 4×10^{-18}, about 2×10^6
10-21 the same, 1.8×10^{23} m⁻³
10-22 (a) (i) 1.0 mol (ii) 2.0 mol
(b) 3.0 mol (c) 40 kPa
10-23 (a) 0.247 mol
(b) (i) $p(8.22 \times 10^{-7}$ mol Pa⁻¹),
(ii) $p(9.69 \times 10^{-7}$ mol Pa⁻¹) (c) 138 kPa
10-24 (a) 1.98 mol (b) 0.17 mol (c) 325 kPa
10-25 (a) 2.0 ms (b) 5.0×10^{-23} N s
(c) 2.5×10^{-20} N (d) 1.25×10^4 N
(e) 104 kPa
10-26 431 m s⁻¹, the same
10-29 (a) 1.9 km s⁻¹ (b) 1.4 km s⁻¹
(c) 0.41 km s⁻¹
10-30 (a) 32.2 m s⁻¹ (b) 32.4 m s⁻¹
10-31 (a) 12.7 g (b) 1.34 km s⁻¹
10-32 (a) 0.48 km s⁻¹ (b) 1.3 g
(c) 2.7×10^{22}
10-33 (a) 273 K (b) 599 K
10-34 all: 6.00×10^{-21} J
10-35 (a) 1.9 km s⁻¹ (b) 0.51 km s⁻¹
(c) 0.21 km s⁻¹
10-36 (a) no (b) yes
10-37 1.5 mm s⁻¹
10-38 50 kPa, 25 kPa, 25 kPa
10-39 (a) 0.93 mol (b) 0.93 kPa
10-40 (a) 100.04 kPa (b) 0.95%, no
10-41 (a) 1.4 mol (b) 0.35 MPa
10-42 1.05
10-43 (b) A (c) 362 m s⁻¹ (d) 226 rev s⁻¹
10-44 (a) (i) 560 (ii) 1200
(b) (i) roughly 63 000
(ii) roughly 136 000
10-45 (a) 7.47 kJ (b) 14.9 kJ, 1200 K
10-46 (a) 28 kJ (b) 40 K
10-47 20.52 J K⁻¹ mol⁻¹,
20.87 J K⁻¹ mol⁻¹
10-49 (a) −0.72 kJ (b) +0.25 kJ
10-50 (a) +0.72 kJ (b) −0.25 kJ
10-52 (a) 0.50 kN (b) 2.5 J
10-53 (a) 0.72 kJ (b) 0.25 kJ
10-54 (a) 6.2 kJ (b) the same (c) the same
(d) only in (a), 6.2 kJ
10-55 (a) w.d. on gas, surroundings heat gas (b) +0.75 kJ (c) +120 K
(d) -6.3×10^{-3} m³
10-56 (a) 100 J
(b) 5.87×10^{-3} m², 6.52×10^{-3} m³
(c) 66 J (d) 166 J
10-57 21 J K⁻¹ mol⁻¹
10-59 (a) 1.18×10^{-2} m³
(b) 1.35×10^{-2} m³
(c) +241 J, −161 J, +80 J

(d) $4.2 \text{ J mol}^{-1} \text{ K}^{-1}$

10-60 (a) 436 J (b) +261 J
(c) *by* the gas, 175 J
10-61 166 J
10-62 0.20 kJ (a) 0.20 kJ (b) 0
10-63 (a) 0.26 kJ (b) 0 (c) 0.26 kJ
10-64 (a) 900 K
(b) ΔU: −750 J, +1500 J, +2250 J,
−3000 J. ΔW: −500 J, 0, +1500 J, 0.
ΔQ: −1250 J, +1500 J, +3750 J,
−3000 J.
(c) +1000 J (d) +5250 J (e) 0.19
10-65 (b) 2.0 (c) 90 kJ (d) 70 kJ
(e) (i) +50 kJ (ii) +20 kJ (f) −85 J
(g) net w.d. by gas in one cycle
(h) −30 kJ

10-67

	ΔW	ΔU	ΔQ	ΔT
(a) (i)	+	0	+	0
(a) (ii)	−	−	−	−
(a) (iii)	0	+	+	+
(b)	+	0	+	0

10-69

	ΔW	ΔU	ΔQ	ΔT
(a) (i)	−	+	0	+
(a) (ii)	+	0	+	0
(a) (iii)	0	−	−	−
(b)	+	0	+	0

10-70 (a) 361 K (b) 57 kPa (c) 274 K
(d) 0 (e) −4.5 kJ (f) 4.5 kJ
10-71 −63°C, higher

11 Real gases and thermodynamics

11-6 8.5×10^{-3} mol
11-7 (b) 518 kPa
11-9 between $2.3 \times 10^{-4} \text{ m}^3 \text{ mol}^{-1}$ and
$8.0 \times 10^{-4} \text{ m}^3 \text{ mol}^{-1}$
11-12 (a) p_o (b) $0.9 N_o$
11-13 (a) (i) halved (ii) doubled
(b) (i) no change (ii) increased
11-14 (a) 1.12 kPa (b) 9°C
11-16 0.37
11-17 (a) 94.0 kPa (b) 64.0 kPa (c) 37.0 kPa
11-21 (a) 35°C (b) − 3.5°C
11-22 177 kPa, no
11-24 109 kPa, lower
11-25 0.54 MW
11-26 (a) 0.80 kJ, 0.10 kJ, 0.30 kJ, 2.4 kJ
(b) −1.05 kJ, +0.30 kJ, +3.15 kJ,
−2.4 kJ
(c) −0.70 kJ, 0, +2.1 kJ, 0
(d) −1.75 kJ, +0.30 kJ, +5.25 kJ,
−2.4 kJ
(e) (i) 4.15 kJ (ii) 5.55 kJ (iii) 1.4 kJ
(iv) 0.25
11-27 (a) +90 kJ (b) 0.50
(c) −45 kJ, −45 kJ
(d) (i) 50 kJ, 100 kJ, 100 kJ, (ii) 0, 0
(iii) −50 kJ, +50 kJ
11-28 (a) 35 J (b) 139 J (c) 0.25
(d) $T_d = 400 \text{ K}$, $\eta_r = 0.25$
11-29 (b) 0.62, 0.42
11-30 (a) 15.0 W (b) 17.4 W (c) 2.4 W
(d) 9.0 W (e) 0.85 W (f) 0.06
11-31 (a) 6.5 kW (b) 2.4 (c) 5.0

11-33 373.16 K, the Carnot heat engine
11-34 (a) (i) 7.0 J K^{-1}, (ii) 5.6 J K^{-1}
(b) 0.24 kJ K^{-1} (c) 2.4 kJ K^{-1}
11-35 (b), (c), (f)
11-36 (a) decreases
(b) (i) 1.41 kJ K^{-1} (ii) 1.21 kJ K^{-1}
11-37 (a) (i) negative (ii) zero
(b) (i) zero (ii) positive
(c) (i) negative (ii) negative
(d) (i) negative (ii) zero
11-38 (a) -113 J K^{-1} (b) greater (c) gain
11-39 (d) (i) $+0.15 \text{ kJ K}^{-1}$,
(ii) -0.15 kJ K^{-1}
11-40 (a) $+352 \text{ J K}^{-1}$ (b) 114 kJ
(c) (i) -363 J K^{-1}, (ii) -352 J K^{-1}
(iii) -341 J K^{-1}
(d) (i) -11 J K^{-1} (ii) 0 (iii) $+11 \text{ J K}^{-1}$
below this temperature $\Delta S_{\text{total}} < 0$, so
the gas cannot dissociate

12 Centripetal forces and gravitation

12-1 (a) $5.2 \times 10^2 \text{ rad s}^{-1}$
(b) $1.7 \times 10^{-3} \text{ rad s}^{-1}$
(c) $7.3 \times 10^{-5} \text{ rad s}^{-1}$
12-2 (a) $7.3 \times 10^{-5} \text{ rad s}^{-1}$
(b) 3.1 km s^{-1}
12-3 (a) 3.5 rad s^{-1} (b) 0.52 m s^{-1}
(c) 0.26 m s^{-1}
12-4 2.1 s
12-5 (a) 100 s^{-1}
(b) $6.3 \times 10^2 \text{ rad s}^{-1}$ 300 s^{-1}
12-6 (a) 250 rad s^{-1} (b) 7.5×10^3 rad
(c) 2.0 rad s^{-2}
12-7 (a) $1.3 \times 10^2 \text{ rad s}^{-1}$
(b) $1.0 \times 10^2 \text{ rad s}^{-1}$ (c) 75 rad (d) 12
12-8 (a) 12 rad s^{-1} (b) 3.0 s (c) 4.0 rad s^{-2}
12-10 (a) $r\theta$ (b) $r\theta/v$
(c) $(2v^2 \sin\frac{1}{2}\theta)/r\theta$
$\theta = 60°$: $a = 0.9549v^2/r$
12-12 (a) 4.0 m s^{-2} (b) 1.0 m s^{-2}
12-13 (a) 463 m s^{-1}
(b) $3.37 \times 10^{-2} \text{ m s}^{-2}$ towards Earth's
centre
12-14 (a) 0.83 m s^{-1} (b) 0.57 m s^{-2}
12-15 (a) 3.1 m s^{-2} (b) 0.28 kN
12-16 (b) 3.0 kN (c) 22 m s^{-1}
12-17 (c) $\tan\theta = v^2/gr$
12-18 (b) 38 m s^{-1} (c) opposite
12-21 (a) (i) zero (ii) not zero
(b) (i) constant (ii) changing
12-22 $3.3 \times 10\text{if-030}^{-8}$ N, the same
12-23 (a) (i) 2.0×10^{20} N (ii) 23 N
(iii) 0.50 kN
(b) (i) $2.8 \times 10^{-3} \text{ m s}^{-2}$ (ii) 0.23 m s^{-2}
(iii) 6.2 m s^{-2}
12-24 (a) 2.2×10^{-7} N (b) 3.5×10^{-7} N
12-25 6.76×10^{-6} N
12-26 (a) 1.9×10^{-44} N (b) 1.2×10^{36}
12-27 7.4×10^{22} kg
12-28 (a) 49 N (b) 5.97×10^{24} kg
(c) $5.5 \times 10^3 \text{ kg m}^{-3}$
12-29 (a) 47 J (b) 16 J kg^{-1}
12-30 (a) 10 m (b) (i) 19.6 N (ii) 19.6 N
(c) (i) 588 J (ii) 196 J (d) 392 J

(e) (i) 20 m s^{-1} (ii) 24 m s^{-1}
12-31 $2.7 \times 10^2 \text{ N kg}^{-1}$
12-32 (a) 2.6×10^6 m (b) 6.4×10^6 m
12-33 (a) (i) $-4.0 \times 10^7 \text{ J kg}^{-1}$
(ii) $-2.0 \times 10^7 \text{ J kg}^{-1}$
(b) $+3.0 \times 10^{10}$ J (c) 7.5 km s^{-1}
12-34 (a) $0.2 \times 10^{-8} \text{ m}^{-1}$, spacecraft
would stop at $r = 5 \times 10^8$ m
(b) 4×10^{10} J (c) 1.7 km s^{-1}
(d) $- 1.8 \times 10^{19}$ J m, $-Gm_E m$,
$4.0 \times 10^{14} \text{ N m}^2 \text{ kg}^{-1}$
12-35 1.5×10^8 J
12-36 (a) 0.44 N kg^{-1} (b) 0.11 N kg^{-1}
12-37 (a) -6.3×10^9 J (b) 6.3×10^9 J
(c) 11 km s^{-1} (d) no (e) 2.4 km s^{-1}
12-39 $V_X + V_Y/10^{-7} \text{ J kg}^{-1} = -3.45$
(at $r = 0.30$ m), -1.56 (at $r = 1.80$ m)
4.8×10^{-10} J
12-40 (a) $3.37 \times 10^{-2} \text{ m s}^{-2}$ (b) 733.5 N
(c) 731.0 N
12-42 84 min
12-43 4.2×10^7 m, 6.6
12-44 1.87×10^{27} kg
12-45 58 km, 1.68 km s^{-1}
12-46 (a) 9.25 N kg^{-1} (b) 9.25 m s^{-2}
(c) 647 N (d) 9.25 m s^{-2} (e) 647 N
(f) no (g) zero

13 Equilibrium and rotation

13-1 (a) $+20 \text{ N m}$ (b) -10 N m
(c) $+7.1 \text{ N m}$ (d) $+7.1 \text{ N m}$ (e) $+11 \text{ N m}$
(f) -35 N m
13-2 front: 120 N, back: 80 N
13-3 (a) 0.27 kN (b) 0.25 kN, 0.10 kN
13-4 (a) 4.0 m from A (b) 9.6 kN, 12 kN
(c) 15.3 kN, 51° above horizontal
13-5 (a) 7.36° (b) 151 N (c) 19.4 N
13-6 40°
13-7 (i) 105 N, 61 N
(ii) $X = T = 12$ N, $Y = 22$ N
13-8 (ii) 22.6° (iii) 0.18 kN
(iv) 0.12 kN, 54° above horizontal
13-9 1.55 kN, 1.77 kN
13-10 1.8 kN
13-11 (a) all: 100 N m clockwise
13-12 (a) $+0.60 \text{ N m}$, -0.20 N m,
$+0.80 \text{ N m}$, -1.20 N m, zero, yes
(b) (i) -0.60 N m (ii) $+0.60 \text{ N m}$ (c) yes
13-13 one author obtained 70 N mm
13-14 (a) 88 N (b) 9.2 N m
13-15 40 N m, 55 N
13-16 9.4 N m, 54 N
13-17 (a) 0.22 kN m (b) 69 kW
13-18 (a) 134 N m (b) 56 kW
13-19 13 N m, 71 N
13-20 17 N m
13-21 (a) 0.28 kN (b) 15 N m
13-22 $\frac{1}{2}ma^2$
13-23 (a) 3.4 kg m^2 (b) 27 kg m^2
13-24 (a) $\frac{1}{3}ml^2$ (c) $\frac{1}{3}ml^2$
13-25 (a) 0.59 kJ (b) 2.4 kJ, 4.2 m s^{-1}
13-26 (a) 2.6×10^{29} J (b) 2.6×10^{33} J
13-27 (a) 1.7 rad s^{-1} (b) 8.7 kg m^2
13-28 (a) 6.1 rad s^{-1} (b) 4.8 m s^{-1}
13-29 (a) 66 mm s^{-1} (b) 4.4 rad s^{-1}

(c) 0.85 J (d) 8.7×10^{-2} kg m^2
13-30 (a) 180 rad s^{-1} (b) 360 rad s^{-1}
(c) 360 rad (d) 57 rev
13-31 (a) -40 rad s^{-2} (b) -80 N m
13-32 B
13-33 5.6×10^{-2} N m
13-34 (a) 7.1×10^{33} N m s
(b) 2.7×10^{40} N m s
13-35 (a) 15 rad s^{-1} (b) 3
13-36 (a) 0.012 kg m^2 (b) 0.40 J
13-37 30.3 km s^{-1}
13-38 1.4 rad s^{-1} anticlockwise,
4.5 m s^{-1}

14 Storing electric charge

14-4 $\pm 1.2 \times 10^{-4}$ C
14-5 2.5×10^2 V
14-6 4.8×10^{-8} F $= 0.048$ µF
14-7 ± 20.0 mC, 5.0 mA, greater
14-8 2.2 µA
14-9 2.0×10^{-4} F
14-10 (a) 0.60 s (b) 18 V
(c) 30 V s^{-1} (d) constant at 14 mA
14-11 (i) (a) 0 (b) 12 V (c) 12 µA
(d) 0 (ii) (a) 12 V (b) 0 (c) 0
(d) $\pm 2.6 \times 10^{-4}$ C; 22 s
14-12 (a) 9.5 V (b) 2.5 V (c) ± 55 µC
14-13 (a) 4.3 V (b) 7.7 V (c) 7.7 µA
14-14 0.54 µA
14-15 4.0 µF
14-17 (a) 11 µF (b) 1.0 µF
14-18 A 18 µC, 6.0 V; B 4.5 µC,
3.0 V; C 13.5 µC, 3.0 V
14-19 3.0 µF in series with 6.0 µF, and
8 µF in parallel with both
14-20 3 groups in parallel, each group
consisting of 3 in series
14-21 7.1 mA
14-22 2.2 µF
14-23 (a) 1.0×10^{11} Ω across input
(b) 1.0 nF across input
14-24 p.d. $\times 0.20$; current $\times 1.0 \times 10^{10}$
power $\times 2.0 \times 10^9$
14-25 (a) 2.5 nC (b) 83 pF (c) 21 pC
(d) 0.83%
14-26 11 pF, 1.1%
14-27 6.0×10^{-10} F, 62.5%
14-28 (a) 5.0×10^{-12} A (b) 1.5×10^{14} Ω
14-29 (a) (i) 850 V (ii) 550 V
(b) 8×10^{-14} A (c) 8×10^{15} Ω
14-30 (a) yes, from about 400 V to 900 V
(b) no (c) (i) about 50 V (ii) about
400 V (d) 70°
14-31 (a) 1.8 J (b) 1.9×10^5 W
14-32 36 V
14-33 ± 2.4 mC, 0.18 J, 100 V, 0.12 J
14-34 (a) 2.0 J (b) 1.0 J
14-35 (a) 3.2 mC (b) 0.96 J
14-36 (a) 4.0×10^{-4} C, 8.0×10^{-4} C
(b) 100 V (c) 4.0×10^{-4} C
(d) 3.0×10^{-2} J
14-37 (a) 0.40 µF (b) 8.0×10^{-5} C
(c) 40 V, 160 V
(d) 1.6×10^{-3} J, 6.4×10^{-3} J
14-38 1.9 µF, 2.1 µF, 2.9×10^{-4} J

14-39 7.9 K
14-40 (a) 1.5×10^2 V (b) 15 µA
(c) 5.6×10^{-2} J (d) yes
14-41 69 s
14-42 (a) 0.67 µA (b) 21 mV s^{-1}
(c) 32 µF (d) 48 s (e) 32 µF
14-43 (a) 15 V (b) t = 35 s (e) 1.8 mC
(f) 7.5 V (g) 240 µF
14-44 (a) 433 s (b) 4.33 MΩ
14-45 (b) 10 V (c) 10 mA (d) 0.50 µA
14-46 (a) 400 µA (b) 4.0×10^{-2} W
(c) 0.80 K s^{-1} (d) 315 K

15 Electric fields

15-5 (a) 2.5×10^{-3} N horizontally
(b) 2.9×10^{-3} N vertically (c) 40°
15-7 -1.4 nC
15-8 1.05×10^{-7} m
15-9 $+2.8 \times 10^{-5}$ J
15-10 (a) 9.6×10^{-18} J (b) 4.6×10^6 m s^{-1}
15-11 (a) 5.0×10^4 V m^{-1}
(b) (i) 1.0×10^{-6} N (ii) 1.0×10^{-6} N
15-12 (a) 2.0×10^4 V m^{-1}
(b) 3.2×10^{-15} N (c) 2.1×10^{-9} s
15-13 6.0 kV m^{-1} B to A
12 kV m^{-1} C to B, 18 kV m^{-1} C to D
zero inside the metal
15-14 (a) 1.0×10^{-8} J (b) 0
(c) 1.0×10^{-8} J
B 600 V, C 200 V, D 600 V
lines of force AB, CD
equipotentials AC, BD
earthed plate 10 mm from AC
other plate 20 mm from BD
15-15 (a) 10 MeV (b) 10 MeV (c) 20 MeV
15-16 (a) 1:2 (b) 1.41:1
15-19 6.4×10^{-19} C
15-20 490 V, top
15-21 (a) $+1.57 \times 10^{-16}$ J
(b) -1.54×10^{-16} J (c) -3.4×10^{-18} J
15-22 -2.7×10^{-9} C m^{-2}, -8.0×10^2 C
15-23 1.8 MV m^{-1}
15-24 (a) 3.3×10^5 V m^{-1}
(b) $\pm 2.9 \times 10^{-6}$ C m^{-2}
(c) $\pm 1.2 \times 10^{-7}$ C (d) $\pm 2.9 \times 10^{-7}$ C
15-25 0.50 µF, 0.62 J
15-26 1.7 nF
15-27 (a) 4.0 nF (b) 1.0 nF
15-28 (a) 2.5×10^{-5} m (b) 0.94 m^2
(c) 48 kJ m^{-3}
length $\times$ width $= 0.94$ m^2
15-29 (a) 35 pF (b) 1.8×10^{-7} C
(c) 4.4×10^{-4} J (d) 10 kV
(e) 8.8×10^{-4} J
15-30 22 pF
15-31 (a) $\pm 9.60 \times 10^{-9}$ C
(b) 1.15×10^{-7} J (c) $\pm 3.20 \times 10^{-9}$ C
(d) 6.40×10^{-9} C charging
(e) $+1.54 \times 10^{-7}$ J (f) -0.77×10^{-7} J
(g) 0.77×10^{-7} J (h) 2.6×10^{-5} N
15-32 (a) 53 s (b) 32 V
15-33 (a) 4.8 µA (b) 3.8×10^{-4} W
(c) 1.9×10^{-4} W
15-35 9.2×10^{-12} F m^{-1}
15-36 1.2×10^5 V m^{-1}
(a) 1.2×10^5 V m^{-1} (b) 5.0×10^4 V m^{-1}

(c) 0.50 kV (d) 0.60 kV (e) 1.7 kV
15-37 (a) 2.7×10^{-5} C m^{-2}
(b) 2.7×10^{-5} C m^{-2}
15-38 4.5 nC
15-44 1.8×10^{-9} N attraction, 3.2 kV m^{-1}
15-45 9.7 nC; yes
15-46 (a) 3×10^{20} V m^{-1} (b) 7 MV
(c) 1×10^2 N (d) 2×10^{-12} J
15-47 4.0×10^{-7} J, 1.0×10^{-5} N
15-48 3.3×10^{-10} F, 4.2 kJ
15-49 (a) 1.5×10^5 V (b) 2.5 µC
(c) 1.0 MV m^{-1}
15-50 1.1 µC s^{-1}, 1.1 µA
15-51 8.8×10^{-12} F m^{-1}
15-52 $E = Q/2\pi \epsilon_o rl$; 1.8 MV m^{-1}

16 Magnetic forces

16-1 (a) turns N pole north
(b) moves also northwards
16-3 (b) zero at some point
(c) increases towards the top
16-6 7.5×10^{-2} N vertically
16-7 OA OB OC 45 mN upwards
OD zero
16-8 (a) 96 N (b) 0.96 kW
16-9 15 mT
16-10 (a) W (b) N (c) N 30° W; 10 µT
16-11 (a) 0 (b) 49°
16-12 (a) 18 µT (b) 49 µT
16-13 8.0×10^{-3} N m
(a) 4.0×10^{-3} N m (b) 0
16-14 0.20 A m^2
16-15 3.0×10^{-4} N
(a) 4.5×10^{-6} N m (b) 4.5×10^{-6} N m
16-16 3.5×10^{-8} N m deg^{-1} or
2.0×10^{-6} N m rad^{-1}
16-17 0.73 T
16-18 (a) 1:4 (b) 5:1
16-20 1.95 µA div^{-1}
16-21 4.8×10^{-14} N
16-22 9.6×10^{-12} N
1.5×10^{15} m s^{-2}
16-23 (a) 1.9×10^{-15} N upwards
(b) 1.2×10^{-15} N upwards (c) 0
16-24 (a) 0 (b) 5.0×10^{-15} N
16-25 (a) 0.15 V (b) 0
16-26 (a) 4.0×10^{-4} m s^{-1}
(b) 9.5×10^{-23} N
(c) 6.0×10^{-4} V m^{-1} (d) 6.0×10^{-7} V
16-27 3.1×10^{22}, 3.0 N; 3.0 N
16-28 component parallel to axis $= 10$ mT
16-29 2.4 mT
16-30 7.0 A
16-31 50 µT; 68 µT and 32 µT
16-32 40° from north
16-33 0.25 T m^{-1}
16-34 9.0 A, E of the wire, NE
16-35 (a) 24 µT (b) 60 µN
(c) F $= (\mu_o I_1 I_2)/(2\pi d)$
16-36 7.0×10^{-3} N repulsion
16-37 (a) (i) $B_X = B_Y = \frac{1}{2} B_Z$
(ii) $P_X = P_Y = \frac{1}{4} P_Z$
(b) (i) $E_X = E_Y = \frac{1}{2} E_Z$
(ii) $W_X = W_Y = \frac{1}{4} W_Z$
16-39 (a) 1.69×10^4 m^2 s^{-1}
(b) 0.85 V left (port) positive (c) no

16-40 4.0 mV, no
16-41 (a) (i) +10 mV (ii) + 2.5 mV
(b) (i) 7.5 mV (ii) 0 (iii) 0 (c) no
16-42 (a) 51 μV (b) 0 (c) 0
16-43 (a) 1.9 A (b) 1.4 W
16-44 17 V, 0.81 N m
16-46 0.637 N m
16-47 (a) 21.8 V (b) 0.40 Ω
16-48 (a) battery 51 W; ammeter,
very small; motor 96 W; resistor 48 W
(b) battery 0.8 W; ammeter, very small;
motor 24 W; resistor 48 W
16-49 (b) 0.80 T (c) 31 V (d) 32 V
16-50 (a) 1.5 A (b) 5.0 A (c) 21 V
(d) 0.11 kW (e) 0.16 kW (f) 67%
16-51 1.2 A, 2.4 A
16-52 (a) speed falls, current rises in
armature (b) no change
16-53 about 2 Ω

17 Changing magnetic fields

17-1 (a) 15 mm right (b) 60 mm left
(c) 60 mm left (c) 30 mm left
17-2 (a) anticlockwise (b) clockwise
(c) clockwise; current ∝ frequency
17-3 (i) (a) × 0.50 (b) × 0.25
(ii) (a) × 1.0 (b) × 0.50
(iii) (a) × 2.0 (b) × 2.0
17-4 (b) 0.30 V (c) 0.26 V (d) 0.16 V
17-5 (a) 63 μV (b) 6.3 mA
17-6 (a) 2.3×10^{-4} V (b) 1.0×10^{-7} W
(c) 0
17-7 45 μT
17-8 (a) 80 mm (b) 20 mm (c) 20 mm
(d) 20 mm (e) 20 mm
17-9 (a) 2.2 V (b) 0.75 V
17-10 8.0×10^{-5} T, 13 mV
17-11 1.5×10^{7} Wb
17-12 1.6 T
17-13 2.5 mT, 1.0 μWb
17-14 0.14 T
17-15 0.15 mH
17-16 0.25 H
17-17 (a) 80 mH (b) 4.0×10^{2} V
17-18 1.6 s
17-19 A to B 52 V; B to C −1.6 kV
17-20 (a) 80 A s^{-1} (b) 0.16 A (c) 2.0 ms
(d) 1.6 mA
17-21 10 V
17-23 (a) 0.30 Wb (b) 94 V
17-24 (a) > 500 Hz (approx)
(b) < 100 Hz (approx); 22 mH, 30 Ω
17-25 1.8 J
17-26 (a) 1.5 A s^{-1} (b) 3.0 A (c) 2.4 kV
(e) 36 J
17-27 < 39 mH
17-28 (a) 10 H (b) 80 J
(c) 2.0×10^{-2} m^{3}
17-29 secondary about 420 turns
(a) current falls (b) no change
17-30 (a) < 3.0 A (b) > 75 mA
17-32 (a) 20 mH (b) 12 (c) 2.9 H
17-33 (a) 30 W and 24 W
(b) 33 W and 20 W
17-34 (a) 60 W (b) 50 W (c) 83%

(d) primary 4.4 W, secondary 2.5 W
(e) 3.1 W
17-38 7.2×10^{2}
17-39 (a) 7.5 H (b) 17 H (c) 47 H

18 Electrons and quanta

18-2 (a) 1.6 mA (b) 26 W
18-3 8.39×10^{2} m s^{-1}
18-4 (a) 9.6×10^{-20} J
(b) 4.6×10^{5} m s^{-1}
18-5 (a) 4.2×10^{6} m s^{-1} (b) 19 ns
(c) 1.6×10^{17} s^{-1} (d) 3.0×10^{9}
18-6 (a) 68 V (b) 164 mA (c) 8.1×10^{2} Ω
(d) < 80 V (e) 1.6
18-7 20 mT
18-8 0.68 T
18-10 2.3×10^{7} m s^{-1}
1.7×10^{11} C kg^{-1}
18-11 (a) $r \times 0.5$ (b) $r \times \sqrt{2}$
18-12 parabolic path (b) 3.5×10^{14} m s^{-2}
(c) 7.5×10^{-9} s (d) 0.16 m
18-13 20 ns
18-14 (a) electrical p.e. (b) k.e.
(c) internal energy, light, X-rays
18-15 (a) 5 oscillations
(b) 2 half-oscillations superimposed;
straight diagonal line
18-16 (a) 2.67 ms (b) 375 Hz
18-17 (a) 1.3 V (b) 125 Hz (c) 2.8 V
(d) 41 mA, 15 mA
18-18 (a) 3.6 kHz (b) 4.3 kV s^{-1}
(c) 60 kHz (d) 1.2 V
18-19 (a) AB +0.20 kV s^{-1}
BC −2.0 kV s^{-1}
(b) AB 44 μA, BC 0.44 mA
18-20 (a) 4.0 ns (b) 1.1×10^{15} m s^{-2}
(c) 4.6×10^{6} m s^{-1} (e) 17°
18-21 3.4×10^{-19} J, 1.8×10^{20} s^{-1}
18-22 (a) 2.3 V (b) 0.41 eV
(c) 3.8×10^{5} m s^{-1}
18-23 (a) 1.00×10^{15} Hz (b) 0.81 eV
18-24 plate slightly positive
(about 2 V); no deflection
18-25 0.41 V
18-26 zinc yes, 4.4×10^{5} m s^{-1}
tungsten no
18-27 (b) 4.2×10^{-15} V Hz^{-1}
(c) 6.7×10^{-34} J s
(d) A 3.6×10^{-14} Hz, B 6.3×10^{14} Hz
(e) A 1.5 V, B 2.6 V
18-28 (a) (i) 10 eV (ii) 1.6×10^{-18} J
(b) (i) 7.3×10^{-23} N s (ii)1.7×10^{-24} N s
(c) (i) 9.1×10^{-12} m (ii) 3.9×10^{-10} m
18-29 (a) 4.4×10^{-24} N s
(b) 1.1×10^{-17} J (c) 67 V
18-30 11 kV
18-32 n = 5, W = −0.55 V
n = 6, W = −0.38 V
7.4×10^{-6} m infra-red
18-33 (a) 13.6 eV, 2.19×10^{6} m s^{-1}
(b) 9.14×10^{-8} m
18-34 1.03×10^{-7} m ultraviolet
1.22×10^{-7} m ultraviolet
6.5×10^{-7} m red
18-35 (a) absorb 2.5×10^{-7} m (UV)
(b) absorb 4.4×10^{-7} m violet

(c) emit 5.4×10^{-7} m green
18-36 4.9 V
18-37 (a) 0.20 m (b) 0.44 m
18-38 0.10 T
18-39 (a) 3.8×10^{13} m s^{-2} (b) 0.51 μs
18-40 (a) 84 mm (b) 0.13 μs (c) 7.6 MHz
18-42 (a) both increased
(b) intensity only increased
18-43 1.9×10^{16} s^{-1}, 0.15 kW
18-44 1.4 kg min^{-1}
18-45 (a) 2.5 eV (light)
(b) 4.1 keV (X-rays in TV tube)
(c) 0.62 eV (infra-red, fire)
(d) 8.3×10^{-10} eV (radio waves)
18-46 6.0×10^{16} Hz, 5.0 nm
18-47 3.1×10^{-11} m
18-48 1.02 MV, X-rays

19 Probing the nucleus

19-3 2.2×10^{5} ion pairs, 8.7×0^{-9} A
singly charged, no recombination
19-6 (a) yes (b) only up to 60 mm
(c) (i) yes (ii) no (d) yes
19-8 (b) -1.77×10^{11} C kg^{-1}
19-9 (a) 1.1 mT (b) 4.2 T
19-12 2.5×10^{-12} m
19-13 0.19 MeV
19-14 neodymium-143
19-15 (a) 3.65×10^{-25} kg
(b) 6.64×10^{-27} kg
(c) 216, 84, 3.59×10^{-25} kg
(d) 1.7×10^{7} m s^{-1}, 1.2×10^{-19} N s
(e) 3.2×10^{5} m s^{-1} (f) 0.12 MeV
19-16 (a) 9.00×10^{16} J (b) 5.63×10^{35} eV
(c) 5.12×10^{5} eV (d) 5.12×10^{5} V
19-17 (a) 1.02 MeV (b) 1.21×10^{-12} m
(c) 0.512 MeV
19-18 (a) 4.39×10^{-14} J (b) 0.275 MeV
19-19 31.973 9
19-20 platinum-192: 78p, 114n
resulting nucleus: 76p, 112n
19-21 19p, 21n, 19e; $^{40}_{18}$Ar
19-22 $^{239}_{92}$U, $^{239}_{93}$Np, $^{239}_{94}$Pu, $^{235}_{92}$U
235.043 90
19-23 (a) 2.8×10^{20} (b) 4.7×10^{-4} mol
(c) 0.11 g
19-24 4.28×10^{-12} J
(a) 2.0×10^{26} atoms (b) 0.34 kg
19-25 170 min^{-1}, 180 min^{-1}
19-26 (a) A 1.25 d, B 3.75 d; ratio = 1:3
(b) 1.9 d; 3:1 (c) 4.8 d; 1:3
19-27 $^{99}_{43}$Tc, $^{99}_{44}$Ru
(a) 6.1×10^{14} (b) 2.2×10^{-13} s^{-1}
(c) 9.9×10^{4} y
19-28 5.7×10^{-11} kg
19-29 (a) 0.25 μCi (b) 1.7×10^{2} d
19-31 9.02×10^{-4} s^{-1}
19-32 2.0×10^{2} litre
19-33 (a) 1.7×10^{-6} mol (b) 1.0×10^{18}
(c) 31.7 s^{-1} (d) 3.1×10^{-17} s^{-1}
(e) 7.1×10^{8} y
19-34 8
19-35 (a) 5.6×10^{-5} (b) 5.5×10^{-5}
19-36 2080 y
19-37 (a) 1.08×10^{11} (b) 3.83×10^{-12} s^{-1}

(c) 347 (d) 1460 y

20 Alternating currents

20-1 2.5 A
20-3 2.8 A, 34 V
20-5 $I_o = 0.884$ A, $f = 50$ Hz
max. power = 300 W
20-6 (a) 18 A (b) 5.8×10^3 A s^{-1}
(c) 1.7 A
20-7 (a) 8.0 kV (b) 2.0% (c) 5.7×10^5 V
20-8 (a) 10 V (b) 7.1 V (c) 71 mA
20-9 (a) 6.37 mC (b) 0.637 A
20.10 both values = V_o; 0.50 kHz
20-13 90° (a) 25 V (b) 35 V
20-14 -9.4×10^4 V s^{-1} (a) 7.0 ms (b)
70%
20-17 (a) $I \propto f$ (b) $I \propto C$
20-18 1.2 A, 340 V
20-19 1.7×10^{-4} F
20-20 (a) $I \propto 1/f$; $I \propto 1/L$
20-21 0.38 H
20-22 sinusoidal, same frequancy,
$V_o = 1.0$ V, in phase with p.d. across
inductor
40 Ω, 60 Ω, 20 Ω
20-23 (a) 40 Ω (b) 6.4 mH
(c) 4.0 μF; increase f:
inductor, current falls
capacitor, current rises
20-24 1.9 H
20-25 (a) 30 Ω (b) 43 Ω (c) 0.32 kΩ
20-26 (a) 50 Ω (b) 80 Ω (c) 62 Ω (d) 0.20 H
(e) 51° current lagging
20-27 (a) 10 Ω (b) 16 Ω (c) 0.77 A
(d) 7.1 W
20-28 (a) 0.81 kΩ (b) 0.30 A
(c) 52° current leading

20-29

	200 Hz	40 Hz	1000 Hz
(a)	16 kΩ	80 kΩ	3.2 kΩ
(b)	23 kΩ	81 kΩ	16 kΩ
(c)	44 μA	12 μA	61 μA
(d)	0.71 V	0.20 V	0.98 V
(e)	0.71 V	0.98 V	0.20 V

20-30 52.0 Ω, 0.382 H, 35.4 μF
20-31 (a) 4.37 kΩ (b) 77.6 mA (c) 233 V
(d) 0 (e) 233 V
20-33 (a) velocity of mass (b) k.e. of mass
(c) e.p.e.
(d) internal energy from friction
20-34 2.9×10^{-10} F
20-35 1650 kHz, 500 kHz
20-36 16 kHz, 100 Ω, 10 V
20-37 2.1×10^2 m
20-38 (a) 5.0 km (b) 5.0 m (c) 5.0 mm
20-39 (a) 1.7×10^7 m s^{-1} (b) 2.25

21 Electronic devices

21-3 (a) 0.6 V (b) 1.4 V (c) 0.35 kΩ
(d) (18.4 V)
21-4 (a) 15 Ω (b) 30 Ω
$E \approx 0.6$ V, $R \approx 6$ Ω
21-5 (a) 0.6 V (b) 63 mA (c) 1.5 V
21-6 square wave, peak value 0.6 V
21-11 (a) 8.5 V (b) 100 Hz (c) 47 mA
(d) 0.94 V (approx) (e) 8.0 V
(f) 17 V

21-13 8.6 kΩ
21-14 (a) 7.6 kΩ (b) 2.5 mA (c) 25 μA
(d) 4.3 V, 63 μA (e) 38 A, 0.7 V
(f) 18 kΩ
21-15 (a) 13 μA (b) 2.6 mA (c) 2.4 V
(d) +10 μA (e) −14 μA
21-16 (a) non-inverting
(c) current and power
(d) $V_{out} = 4.3$ V, $I_e = 10$ mA
$I_b = 0.20$ mA (e) $V_{in} = 0.5$ V
(f) 0.96 (i.e. 43/45)
21-17 $I = 0.37$ A, max. $R_L \approx 16$ Ω
the transistor bottoms
21-18 (a) −100 (b) +1.0 (c) +11
21-20 9.0 V, inverting amplifier
21-22 (a) high (≈ +5 V) (b) 4.3 V
(c) 0.50 mA (d) 8.6 kΩ (e) 3.6 V
21-23 (a) OR (b) AND (c) NOR
(d) AND (e) OR
21-24 0000, 0001, 0010, 0011, 0100,
0101, 0110, 0111, 1000, 1001, 0000
21-25 (a) 1.0 V (b) 3.0 V (c) 7.0 V

22 Oscillatory motion

22-5 (a) 8.6×10^4 s, 1.2×10^{-5} Hz
(b) 0.83 s, 1.2 Hz (c) 0.020 s, 50 Hz
(d) 3.9×10^{-3} s, 260 Hz
22-6 0.42 m s^{-1}
22-7 170 m s^{-1}
22-8 s^{-2}; $f_A = \sqrt{10} f_B$ but no information
about amplitude
22-9 (a) 1.6 m (b) 0.67 s (c) 1.6 m
(d) zero (e) −140 m s^{-2} (f) 110 N
22-10 0.60 s, 25 ms
22-11 2.05 p.m.
22-12 2
22-13 $\sqrt{2}$; 2
22-14 (a) 0.32 m s^{-2} (b) 1.6 s or $\pi/2$ s
(c) 0.080 m s^{-1}
22-15 4.4 kN, 88 J
22-16 930.3 mm
22-17 27 flashes for half an oscillation
⇒ $l = 0.80$ m
22-18 5 swings and 6 swings
22-19 1.6 s
22-20 7.1°
22-21 (c) $E \propto y_o^2$
22-23 (c) 4.0 m s^{-1}, 3.5 m s^{-1}
22-27 13 m s^{-1}
22-28 (a) 0.74 m (b) 200 N m^{-1}

23 Describing waves

23-5 42 g
23-6 half way down $c = 6.3$ m s^{-1},
1 m from end $c = 3.2$ m s^{-1}
23-7 (b) 6.3 m s^{-1}
23-8 5200 m s^{-1}
23-9 (b) 0.1 m s^{-1} to 100 m s^{-1}
(c) 4 km s^{-1} to 14 km s^{-1}
23-10 $c = k\sqrt{(E/\rho)}$; $k = 0.99 \approx 1.0$
23-11 24 N m^{-1}
23-12 620 km
23-13 (a) (i) 2.5 s, 4.7 s (ii) 3.3 s, 3.9 s
23-14 $c = \sqrt{(\gamma/\rho\lambda)}$
23-18 All k.e. in the moving rope

23-19 (a) 0.60 m (b) 6.0 s
(c) 0.17 Hz
23-20 yes, for seagulls $3\lambda/2$ apart on a
line perpendicular to the waves
23-21 (a) 0.25 m (b) 1.5 m (c) 6.8 m s^{-1}
(d) 4.5 Hz (e) 2.2 s
23-22 (a) 0.63 m (b) 40 Hz (c) 5.0 m s^{-1}
23-23 (a) 16 W (b) 9.0 W
23-26 $n \times 2.28$ Hz, $n = 1,2,3$, etc.
23-27 (a) 333 Hz (b) 500 Hz
(c) 667 Hz; $n \times 333$ Hz where $n = 1,2,3$,
etc.
23-30 5110 m s^{-1}
23-31 (b) $1/4l^2\mu$ (d) $2\sqrt{(\mu/T)}$
23-32 $T \propto \mu$
23-33 53 g
23-34 52 N
23-35 (a) 3.8 m s^{-1} (b) 53 mm

24 Sound

24-1 10π rad
24-6 (a) $S_2M = 0.80$ m $\pm n(0.08$ m)
(b) as (a) plus 0.04 m
24-8 320 m s^{-1}
24-10 2.5×10^{-9} J
24-11 98.2 N
24-12 13°
24-13 6 mm to 15 mm
24-14 340 m s^{-1}; 0.58 m
24-15 720 Hz
24-16 $c = k\sqrt{(p/\rho)}$; c is independent of λ
24-17 0.71 s
24-18 (a) 10 mm (b) 14 mm
24-21 74° (semi-angle 37°)
24-23 maxima heard every 1.0 s
24-24 (a) nf_A, nf_B, $(2n − 1)f_C$ where $n =$
1,2,3 etc. (b) 30 Hz, 15 Hz
24-25 340 m s^{-1}; end correction 20 mm
24-26 570 Hz, 1130 Hz, 1700 Hz
24-28 (a) 3.5 kHz (c) 340 m s^{-1} (d) $1.0 \times$
10^{11} Pa
24-29 0.83 m
24-30 lowest note is 760 Hz
24-31 (a) 9.9×10^{-7} W m^{-2}
(b) 0.3×10^{-7} W m^{-2}; amplitude $\propto 1/r$

25 Wavefronts and rays

25-2 (a) 8980 W m^{-2} (b) 50 W m^{-2}
25-3 (a) (i) 8.5 J m^{-1} (ii) 4.2 J m^{-1}
(b) 35 mm
25-5 60°; 30°
25-7 $\theta_a = 59°$; angle of ray to
perpendicular in oil is 36°
25-8 (a) 0.10 m s^{-1} (b) 29°
25-9 46 ns
25-11 27°; 27 mm
25-12 19° 51′; 3′; no
25-13 (a) 48.5° (b) 1.62
25-14 31°, 75°; the ray is symmetric
25-15 For $\theta_p = 20°$ angle of refraction
= 35°, angle of reflection = 20°, etc.
25-16 0.49° or 29′
25-17 ray emerges at 29° to the water
surface

25-18 738 nm, 472 nm
25-19 53°
25-20 for $A = 20°$, $D = 10.9°$ etc.; 0.50; yes
25-21 2.1 m
25-22 1.48
25-27 0.80 m
25-31 (i) right (ii) 1.5 (iii) 1.5
25-32 2.26 m
25-33 (a) (i) 4.04° (ii) 3.86° (b) 0.18°
25-34 15.0°; 3.86°

26 Optical systems

26-1 (a) R, 0.25 m, 1.3 mm, I
   (b) R, 0.40 m, 5.0 mm, I
   (c) R, 2.2 m, 50 mm, I
   (d) V, 0.60 m, 20 mm, U
26-2 (a) 0.16 m, 0.83 mm
   (b) 0.13 m, 1.7 mm
   (c) 0.10 m, 2.4 mm
   (d) 0.086 m, 2.9 mm; all V and U
26-3 (a) 0.24 m (b) 2.5 mm
26-4 (a) 2.0 m (b) 0.20 m
26-5 $f = 28$ mm, 25 mm
26-7 5.6 mm
26-9 203 mm; ±4 mm or ±2%
26-10 30 mm beyond the diverging lens
26-12 0.31 mm
26-13 R, 67 mm beyond the second lens, × 1.33, I
26-14 (a) R, 0.15 m in front, 0.15 mm, I
   (b) R, 0.30 m in front, 10 mm, I
   (c) V, 0.30 m behind, 30 mm, U
26-15 0.15 m, 0.15 mm (b) 0.10 m 3.3 mm
   (c) 0.060 m, 6.0 mm; all behind mirror
26-17 0.15 m
26-18 1.7
26-19 $f = r/0.39$
26-21 (a) 1.30 m (b) 1.06 m
26-22 diameter red circle on S_b = 11 mm, diameter blue circle on S_r = 14 mm
26-23 0.24 mm etc.
26-24 0.19 mm (a) unchanged
   (b) 9 times brighter
26-25 near point at 250 mm
26-26 about 4.5 mrad, 3×
26-27 86' and 43'
26-28 $m = 0.5$; (a) 80 mrad
   (b) 35 mrad; $M = 2.3$
26-29 yes—just
26-30 (a) f of lens
   (b) aperture/f; 9.0 mrad (0.52°)
26-31 (a) 5×, U (b) no image seen
   (c) 4.9×, I
26-32 (a) 4×, 200 mm from 0
   (b) $OE = 270$ mm (c) $m = 14$
26-33 O:23 mm left of O, R, 0.25 mm; E:200 mm left of E, V, 6.4 mm
26-35 6.6 mm
26-36 at F_o, about 200 mm across
26-37 560×
26-38 (a) 4 times as bright (b) twice as big
26-39 (a) 7.8 mm (b) 0.039 rad (c) 0.33×
26-40 $f/14$, $f/3.2$; 19
26-42 (a) 20 mrad (b) 8.0 m
   (c) 120 mrad in both cases

(d) 6× in both cases
26-43 0.11 m from E, diameter 7.5 mm; no; probably not
26-44 59° 58'
26-46 1.583

27 Patterns of superposition

27-1 0.30 mJ, 300 W
27-2 (a) 144 mm (b) 4.5λ; same
27-3 8 mm (a) maximum (b) minimum
27-4 48 mm
27-6 0.50 m
27-7 800.006 4 mm, 800.004 9 mm, 1.5 μm; minimum
27-9 about 48 mm; yes
27-10 ratio $d/D = 0.34$
27-11 510 Hz
27-14 20λ and 24λ; antiphase
27-15 (a) 50889 (b) 50901; 25π rad
27-16 (a) minimum at 875 m and 2625 m; maximum at 1750 m
27-18 48 μm
27-19 690 nm
27-20 path difference = 16λ, with one phase reversal $\Rightarrow$ minimum
27-21 560 nm
27-22 28 mm
27-23 39.5°
27-24 minima at 24° and 53°
27-25 560 nm; maxima separated by about 0.05°
27-26 (a) 0.28 mm (b) 0.08 mm
27-27 no, beam has a diameter of 0.32 m at the flag
27-29 yes, just
27-30 $2\theta = 4.0°$ (circular aperture)
27-31 (a) 0.29 μm (b) 0.21 μm
27-32 (a) 1.57 μm, 635 lines mm^{-1}
   (b) 646 nm
27-33 the zero order plus seven on each side
27-36 (a) 10.5° (b) 37 mm
27-38 6×10^{-7} m
27-40 1040 nm
27-42 (b) very tall narrow maxima but at the same angular spacing as for two slits
27-43 154 pm; 30.5°, 49.6°
27-44 109 pm

28 The electromagnetic spectrum

28-2 (a) 10^{-18} s (b) 15 ns (c) 1.2 s
28-4 9.5×10^{-25} J or 6 μeV
28-8 2.9×10^8 m s^{-1}
28-9 69 rev s^{-1}
28-10 ± 5.0 s
28-11 2.98×10^8 m s^{-1}
28-12 6.8×10^6 m s^{-1}
28-13 134 Hz; no
28-14 1.6×10^{-26} N
28-15 (a) (i) 0.25 GHz (ii) 2.8 μV
   (b) $\pi/2$ rad
28-16 1.5×10^6
28-17 ghost signal has travelled an

extra 1.5 km
28-18 104 m; fading every 3.5 s
28-21 60 W, assuming an inverse square law
28-22 (a) 2×10^{-13} W m^{-2}
   (b) about 3 mrad
28-23 3.9×10^{26} W; 4.3×10^9 kg s^{-1}
28-26 58°
28-27 109.0, 109.5; π rad
28-28 29 MJ m^{-2}, 13 MJ m^{-2}, 3.3 MJ m^{-2}: about 9 times
28-29 36 mW; 30 °C
28-30 (a) 840 kW (b) 2.1 kW
28-31 0.34 kW m^{-2}, 5.3×10^{-8} W m^{-2} K^{-4}
28-32 58 W
28-33 5.0 s
28-34 1300 K for a black body; too low
28-35 yes; 4.0×10^{-3} m K; tungsten does not emit as a black body
28-36 (a) 1600 W (b) 0 W
   (c) 25 W emitted
28-37 6.6 W
28-38 3.4 W; 52 mK s^{-1}
28-41 (a) 61 mm (b) use a sheet of glass
28-42 6.2×10^{-20} J or 0.39 eV, 2.4×10^{-20} J or 0.15 eV

Appendix: Useful mathematics

1 max: 0.465 m × 0.365 m = 0.169 7 m^2
   min: 0.455 m × 0.355 m = 0.161 5 m^2
   best estimate of area = 0.17 m^2, 17 kN
2 0.006 2 m^3
3 kg m^{-3} (a) 1.29 kg m^{-3}
   (b) 1.36×10^4 kg m^{-3}
   (c) 8.9×10^3 kg m^{-3}
4 $1\,\Omega = 1$ kg m^2 A^{-2} s^{-3}
6 $2.997\,925 \times 10^8$ m s^{-1}
7 (a) 8.0×10^{-10} F
   (b) 1.5×10^3 V (c) 5.0×10^7 W
   (d) 4.0×10^{-8} s (e) 1.2×10^{-7} C
   (f) 2.0 J (g) 1.7×10^7 V
8 60p
9 1.4 kW m^{-2} or 1.4 kJ m^{-2} s^{-1}
10 If $\Delta F = 0$, $\Delta v = a\Delta t$
12 $f \propto 1/\lambda$
   gradient of (b) = 3.4×10^2 m s^{-1}
13 -0.60 V A^{-1}
   $V = 1.52$ V $- (0.60$ V A$^{-1})I$
14 (a) 0.92 s (b) -0.60 kg, 0.47 s^2
   (c) 7.9 s^2 kgif-030-1
   (d) $b = 0.060$ kg, $k = 0.127$ kg s^{-2}
15 1.1 m s^{-2}
16 (a) $\pi/3$ rad = 1.05 rad (b) 2.62 m
17 0.66 m
18 (a) 3.0×10^{-4} rad (b) 1.0 min
19 3.09×10^{16} m, 0.758 sec
20 19 m
21 (a) 44° (b) 42°, error $\approx 5\%$
22 1.4 m s^{-1}, 3.8×10^2 m s^{-2}
23 (a) 3.1×10^6 V s^{-1} (b) 0.63 V
24 2×10^{-5} Hz, 2 mm s^{-1}
26 0.14 h^{-1}, 1.3×10^{10}
27 yes, 2.2 d, after 1.5 d